T0254899

Intuitive Understanding of Kalman Filtering with MATLAB®

Intuitive Understanding of Kalman Filtering with MATLAB®

Armando Barreto

Malek Adjouadi

Francisco R. Ortega

Nonnarit O-larnnithipong

CRC Press is an imprint of the
Taylor & Francis Group, an **informa** business

First edition published 2021
by CRC Press
6000 Broken Sound Parkway NW, Suite 300, Boca Raton, FL 33487-2742

and by CRC Press
2 Park Square, Milton Park, Abingdon, Oxon, OX14 4RN

CRC Press is an imprint of Taylor & Francis Group, LLC

Library of Congress Cataloging-in-Publication Data
A catalog record for this book has been requested

ISBN: 978-0-367-19135-1 (hbk)
ISBN: 978-0-367-19133-7 (pbk)
ISBN: 978-0-429-20065-6 (ebk)

Typeset in Minion Pro
by Apex CoVantage, LLC

Contents

Acknowledgments

THE INTEREST IN THE Kalman Filter that triggered the review of various sources describing it; the reflections about its components; its simulation and real-time implementation was prompted by the participation of the authors in research projects where this kind of estimator can be used advantageously. In particular, the authors wish to acknowledge the support from grants CNS-1532061, IIS-1948254, CNS-1551221, CNS-1338922, BCS-1928502 and CNS-1920182, from the National Science Foundation.

Authors

Armando Barreto, PhD, is Professor of the Electrical and Computer Engineering Department at Florida International University, Miami, as well as the Director of FIU's Digital Signal Processing Laboratory, with more than 25 years of experience teaching DSP to undergraduate and graduate students. He earned his PhD in electrical engineering from the University of Florida, Gainesville. His work has focused on applying DSP techniques to the facilitation of human–computer interactions, particularly for the benefit of individuals with disabilities. He has developed human–computer interfaces based on the processing of signals and has developed a system that adds spatialized sounds to the icons in a computer interface to facilitate access by individuals with "low vision." With his research team, he has explored the use of Magnetic, Angular-Rate and Gravity (MARG) sensor modules and Inertial Measurement Units (IMUs) for human-computer interaction applications. He is a senior member of the Institute of Electrical and Electronics Engineers (IEEE) and the Association for Computing Machinery (ACM).

Malek Adjouadi, PhD, is Ware Professor with the Department of Electrical and Computer Engineering at Florida International University, Miami. He received his PhD from the Electrical Engineering Department at the University of Florida, Gainesville. He is the Founding Director of the Center for Advanced Technology and Education funded by the National Science Foundation. His earlier work on computer vision to help persons with blindness led to his testimony to the U.S. Senate on the committee of Veterans Affairs on the subject of technology to help persons with disabilities. His research interests are in imaging, signal processing and machine learning, with applications in brain research and assistive technology.

Francisco R. Ortega, PhD, is an Assistant Professor at Colorado State University and Director of the Natural User Interaction Lab (NUILAB). Dr. Ortega earned his PhD in Computer Science (CS) in the field of Human-Computer Interaction (HCI) and 3D User Interfaces (3DUI) from Florida International University (FIU). He also held a position of Post-Doc and Visiting Assistant Professor at FIU. His main research area focuses on improving user interaction in 3DUI by (a) eliciting (hand and full-body) gesture and multimodal interactions, (b) developing techniques for multimodal interaction, and (c) developing interactive multimodal recognition systems. His secondary research aims to discover how to increase interest for CS in non-CS entry-level college students via virtual and augmented reality games. His research has resulted in multiple peer-reviewed publications in venues such as ACM ISS, ACM SUI, and IEEE 3DUI, among others. He is the first-author of *Interaction Design for 3D User Interfaces: The World of Modern Input Devices for Research, Applications and Game Development* (CRC Press, 2020).

Nonnarit O-larnnithipong, PhD, is an Instructor at Florida International University. Dr. O-larnnithipong earned his PhD in Electrical Engineering, majoring in Digital Signal Processing from Florida International University (FIU). He also held a position of Post-Doctoral Associate at FIU between January and April 2019. His research has focused on (1) implementing the sensor fusion algorithm to improve orientation measurement using MEMS inertial and magnetic sensors and (2) developing a 3D hand motion tracking system using IMUs and infrared cameras. His research has resulted in multiple peer-reviewed publications in venues such as *HCI-International* and *IEEE Sensors*.

Introduction

THE KALMAN FILTER, ENVISIONED by Dr. Rudolf E. Kalman (1930–2016) provides an effective mechanism to estimate the state of a dynamic system when a model is available to sequentially predict the state and sequential measurements are also available. This is a common kind of situation in the study of many practical dynamical systems, in diverse fields.

The Kalman Filter has had an important impact on advances within the fields related to the navigation of ships, aircraft and spacecraft. In those initial areas of use, the Kalman Filter was almost exclusively used for highly specific applications, by a very small group of highly specialized users. However, in the XXI century its potential application to small robots and miniature unmanned vehicles has broadened the appeal of this powerful estimation approach to a much wider audience. Furthermore, the recent emergence of cheap, miniaturized Inertial Measurement Units (IMUs) or Magnetic, Angular Rate and Gravity (MARG) sensing modules in contemporary applications has magnified the importance of techniques such as Kalman Filtering, capable of combining information from multiple sensors. Of course, all of these contemporary applications work on digitized data, and, as such, are addressed by the *Discrete Kalman Filter*, which is the subject of this book.

This book is our effort to provide that wider audience with a presentation of the Kalman Filter that is not a mere "cookbook" list of steps (which may result in a sub-optimal use of this important tool), while not requiring the reader to wade through several formal proofs to accomplish a strict derivation of the algorithm. Those proofs were given by Kalman in his 1960 paper and others who have studied the issue from a formal perspective since then.

Instead, it is our hope to provide the reader with an explanation of the Kalman Filter approach that requires a minimum of background concepts, presented simply and succinctly in *Part I* of this book. We expect that this concise background review will, nonetheless, be enough to help the reader see why it is that the Kalman Filter is capable of obtaining improved estimates of the variables studied. Most importantly, we would like to help the reader acquire an *intuitive understanding* of the meaning of all the parameters involved in the Kalman Filter algorithm and their interactions. The development of that intuitive understanding of the elements at play in the Kalman Filter, and how they come together, according to the background concepts, is the objective of *Part II* of this book. The chapters in Part II lead the reader to the formulation of the Kalman Filter algorithm as the logical conclusion of the considerations made throughout that part of this book. *Part III* of this book focuses on off-line implementation of the Kalman Filter to address estimation challenges of increasing complexity. This third part starts by leveraging the understanding of the Kalman Filter algorithm to develop a MATLAB® implementation for a single Kalman Filter iteration. The rest of the off-line examples in this book are developed from that basic one-iteration function, which can be tailored to create off-line Kalman Filter solutions to diverse estimation problems. *Part IV* of this book includes two chapters that focus on the application of Kalman Filtering to the estimation of orientation of a miniature IMU module, utilizing the gyroscope and accelerometer signals provided by the module. The first of the two chapters of this part develops an off-line solution that still uses the basic one-iteration MATLAB® function created at the beginning of Part III. The gyroscope and accelerometer signals are read from a pre-recorded file which was written while the IMU module was subjected to a specific series of orientation changes, so that the results can be compared with the known series of orientations experienced by the module. The second chapter in Part IV of this book reproduces the implementation of the algorithm in a C program which calculates the Kalman Filter results in real time and stores them to a text file, for later review.

Our emphasis in writing this book has been to foster in the reader an intuitive understanding of the Kalman Filtering process. To that end, we have tried to use analogies and pictorial representations to inform the reader of the main concepts that need to be understood in order to "make sense" of the elements in the Kalman Filter algorithm, their practical meaning, their relationships and the impact that each one of them has in the fulfillment of the goals of the Kalman Filter. Also to that end, we have included a number of "MATLAB® code" segments, which are sometimes scripts (i.e., plain sequences of MATLAB® commands) and some other times custom MATLAB® functions. In either case, the reader is encouraged to execute the MATLAB® code as originally presented, to observe the results in MATLAB®, *and then explore the concepts by modifying the parameters involved in the code, observing the resulting changes in the outcomes*. This type of exploration of the concepts may be powerful in helping to develop the intuition we want the reader to achieve.

While we, and everyone else, refers to the algorithm that is the subject of this book as a "filter," we would like to warn the reader that the algorithm is really a "state estimator," which, depending on the context, may have multiple signals "coming in" as "inputs" (which we will mainly identify with the "measurements") and yield values of multiple "state variables" after each iteration. Therefore, the Kalman Filter is not constrained to the most familiar framework in which one "input signal" is fed into a "filter" to obtain from it a "cleaner output signal." Nonetheless, some specific configurations of the Kalman Filter estimator may actually involve two signals for which the "cleaning" effect of an input may be apparent in an output signal. We will show a couple of such instances in the examples.

Finally, a few important notes:

- This book focuses on the *Discrete Kalman Filter*. Therefore, with the exception of the initial discussion in Section 1.1, *we will be dealing with discrete-time sequences*. However, to keep compatibility with the notation used in many sources related to Kalman Filtering (including Kalman's 1960 paper) we will denote those sequences as, for example, x(t). In other words, we will use "t" as a *discrete-time* index. (This is in contrast to many signal processing books which use "t" for continuous time and another variable, e.g., "n" to index discrete time.)
- Our discussions will involve scalar variables and matrices (or vectors, as sub-cases of matrices). To differentiate them we will use **bold typeface** for matrices and vectors. For example, $v_1(t)$ represents a *scalar* sequence of (digitized) voltages, while **F**(t) is the state transition *matrix*, which may vary from sampling instant to sampling instant, as it is shown as a function of "t."
- In addition to having included the listings of all the "MATLAB® codes" in the chapter where each is first mentioned, all these files can be retrieved from the repository at https://github.com/NuiLab/IUKF.
- Due to the font and margins used, in some instances, a long line of MATLAB® code may be printed in this book partially "wrapped" into the next line. The files in the repository do not have this effect.

- In the "MATLAB® codes" provided we have used a limited palette of colors for the figures, so that it might still be possible to distinguish several traces in the figures printed using grayscale levels for the physical edition of this book. The readers are invited to change the color assignments to take full advantage of the color capabilities of the computer monitors they will be using when executing the functions in MATLAB®.

The MathWorks, Inc.
3 Apple Hill Drive
Natick, MA 01760-2098 USA
Tel: 508 647 7000
Fax: 508-647-7001
E-mail: info@mathworks.com
Web: www.mathworks.com

I

Background

THE OBJECTIVE OF THE three chapters in this part of the book is to provide the reader with the background concepts that will be essential to understand the elements involved in the kind of estimation challenge that the Kalman Filter addresses. This background is also necessary to follow the reasoning steps that will lead us to the Kalman Filter algorithm, in Part II.

We expect that the reader of this book will be primarily interested in understanding and applying the Kalman Filter. As such, we have tried to keep the background chapters in this part of the book very concise and "to the point." Clearly, any of the concepts outlined in these chapters could be developed much more extensively, and put into a wider, more formal context. Instead we have tried to communicate only the concepts that will be used in our future reasoning (Part II) and we have emphasized descriptions of these concepts in familiar terms, whenever possible.

We strongly suggest that the reader reviews these short background chapters. We hope these chapters will "fill any gaps" in the knowledge of some of our readers, so that they will be well equipped to follow and assimilate our reasoning through Part II. Even if you believe you are fairly familiar with all these background items, it will likely be worthwhile reviewing these chapters to make sure you identify our specific view of these concepts, which should facilitate your reading of Part II.

CHAPTER 1

System Models and Random Variables

This chapter presents the reader with the concept of a model for a system and justifies the need to address some variables as random (and not deterministic) variables. The remainder of this chapter is devoted to providing an intuitive understanding of the typical mechanisms that are used to characterize random variables, emphasizing the understanding of how the characterization applies to the digitized values of a signal. The latter portion of this chapter focuses on the Gaussian (Normal) probability distribution, its notation, the effect of processing samples that have a Gaussian distribution through a transformation represented by a straight line, and the characteristics exhibited by a random variable that is the result of multiplying two variables that have Gaussian probability distributions.

1.1 DETERMINISTIC AND RANDOM MODELS AND VARIABLES

Engineers very often aim at providing real-life solutions to real-life problems. However, engineers very seldom attack the problem directly on the physical reality or circumstance where the "problem" lies. Instead, engineers abstract the critical functional aspects of a real-life problem in an operational model. The model frequently is an intangible representation of the real situation that can, nonetheless, be manipulated to predict what would be the outcomes of different changes in the inputs or the structure of the real-life situation.

Engineers use knowledge of the physical world to emulate real-life constraints in the manipulations of their abstract models, so that results obtained from the models are good predictors of what will happen if the same manipulations applied to the model are implemented in the real world. Those real-life constraints and the manipulations performed in the model are commonly expressed and achieved in mathematical terms. For example, Kirchhoff's Voltage Law states that the net sum of voltages around a loop is zero (Rizzoni 2004):

$$\sum_i V_i = 0 \tag{1.1}$$

Using this and the model of the relationship between the current i_R through a resistor R and the voltage V_R across its terminals ("Ohm's Law"):

$$V_R = R I_R \tag{1.2}$$

one can calculate, using only algebra, the voltages at the nodes of a resistive circuit with one loop.

For example, if we "simulate" a sinusoidal voltage source, $V_s(t) = \sin(\omega t)$, applied to the circuit in Figure 1.1, we can use the model to develop an "equation" that would predict the value of the output voltage $V_o(t)$ for any time t:

$$V_s(t) - V_{R1}(t) - V_{R2}(t) = 0 \tag{1.3}$$

$$V_s(t) - \left(R1 * I_{R1}(t)\right) - \left(R2 * I_{R2}(t)\right) = 0 \tag{1.4}$$

$$I_{R1}(t) = I_{R2}(t) = I(t) \tag{1.5}$$

$$V_s(t) = (R1 + R2) * I(t) \tag{1.6}$$

$$I(t) = \frac{V_s(t)}{R1 + R2} \tag{1.7}$$

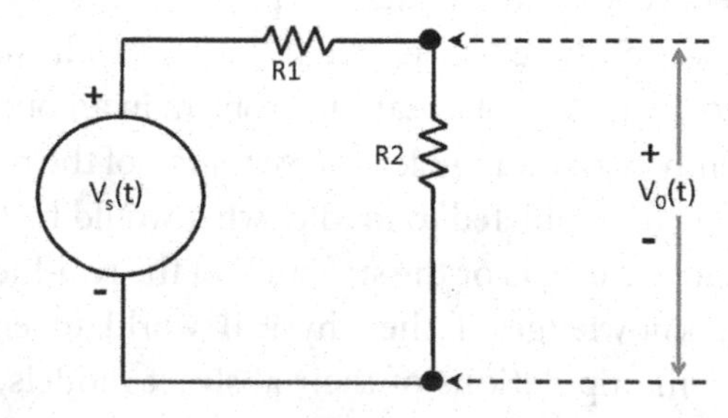

FIGURE 1.1 Simple electrical circuit with just one loop.

And

$$V_o(t) = V_{R2}(t) = R2 * I(t) \tag{1.8}$$

So

$$V_o(t) = V_s(t)\frac{R2}{R1 + R2} \tag{1.9}$$

That is:

$$V_o(t) = \frac{\sin(\omega t) * R2}{(R1 + R2)} \tag{1.10}$$

If we attach to this model "nominal" values of the elements (R1, R2, ω) that replicate the real behavior of the physical components in a real-life circuit, we could "predict" the value of $V_o(t)$ at any desired time t. For example, if R1 = 10,000 Ohm, R2 = 10,000 Ohm and ω = 3.1416 radians per second, then our model predicts:

$$V_o(t) = \frac{\sin(3.1416t) * 10000}{(10000 + 10000)} \tag{1.11}$$

Or

$$V_o(t) = 0.5\sin(3.1416 * t) \quad Volts \tag{1.12}$$

From this we could predict that at t = 0.5 seconds, V_o = 0.5 volts. If t = 1.0 seconds, V_o = 0 V. If t = 1.5 seconds, V_o = −0.5 V, etc.

Equation 1.12 is the analytic expression of a *deterministic* model for $V_o(t)$. You plug in the value of t and the model tells you what the specific value of V_o will be (as estimated by this model). And you get a specific answer for each and any value of t you may want to ask about.

It is important to highlight three aspects of these deterministic models:

1. They will yield a specific value of the output variable for any value of the input variable, "no exceptions," which sounds great, and makes engineers very enthusiastic about them.
2. But, getting "an answer" from these models does not imply in any way that you are getting "the real answer." That is, if the model is not

appropriately matched to the real-life situation, the value of the estimation provided by the model could be completely wrong. (I.e., it could be very different from what happened in the physical system being studied.) As a simple example, consider how wrong the estimates obtained from Equation 1.12 would be if it was used to predict the output voltage of a circuit exactly as described earlier, *but* with a resistor R2 = 1000 Ohms!

3. The deterministic characterization of the signal $V_o(t)$ allows us to fully describe it to another person. I could send an email to a colleague in Germany including Equation 1.12 and she could use it to obtain the same plot of $V_o(t)$ that I can get where I am.

However, we are unable to predict the values of many real-life signals in the definitive way provided by a deterministic characterization, such as Equation 1.12. For example, there is no deterministic model that will accurately tell me the value of voltage coming from a microphone, $V_o(t)$, when I clap my hands next to it.

We know the microphone will yield a voltage that varies through time, alternating above and below 0 V, but there is no equation that can tell me the specific value of voltage that an oscilloscope will show right at t = 0.020 seconds (20 milliseconds), from the moment when my two hands first touch each other when I clap. This is an example of a signal that cannot be easily characterized with a deterministic model. In this example, $V_o(t)$ is best characterized as a "random variable."

This seems to be a considerable setback. How can I characterize a random variable like the voltage signal generated when I clap? Is there anything I can say about it to describe it? Or, in other words, how can I try to describe a random variable? If we were to look at the voltage coming out of a ("dynamic") microphone when I clap, we would notice that there are "certain things" that can be said about it. If we are looking at the voltage straight from the microphone, we will see that it is very unlikely to ever reach outside a (symmetric) range of voltages. For example, it is not likely to be above 1 Volt or below -1 Volt. In fact, it seems that "most of the time" the voltage level is closely above or below 0 Volts.

1.2 HISTOGRAMS AND PROBABILITY FUNCTIONS

Furthermore, there is a "standard way" in which we can summarize these observations about which values are taken by the random signal (particularly well suited for a digitized signal). We can develop the *histogram* of the digitized signal (O'Connell, Orris, and Bowerman 2011; Field 2013).

A histogram for a digitized signal is built simply by dividing a range of amplitudes of the signal in adjacent "bins" of a certain width (for example, one bin from 0.00 to 0.05 and then another bin from 0.05 to 0.10, etc.), and then "placing" each value of the random time series representing the digitized signal in its corresponding bin, according to its amplitude. After doing this for the whole time series, the plot of all the bins, each represented by a rectangle whose height is the number of signal samples ultimately contained in that bin, is the histogram of that signal. Whoever develops the histogram needs to decide first on the location of the bin boundaries. Usually, the width of all the bins is the same. Figure 1.2 shows the creation of a histogram for a random discrete-time signal. (The signal is assumed to have many more than the five samples shown in the figure to illustrate the process.)

Important Note: Please be aware that, from this point forward, we will use "t" as a discrete-time index (McClellan, Schafer, and Yoder 2003, 2016). That is, we will use "t" when referring to discrete samples of a digitized signal, such as x(t). We have chosen to use "t" to represent discrete time to keep compatibility with other texts discussing Kalman Filtering, including Kalman's paper (Kalman 1960).

The histogram is one "practical" tool that we have for the characterization of a random signal. We may not be able to "fully" ("deterministically") forecast the value of the signal at any arbitrary time, but we can at least convey to others which amplitude values the random signal takes on more frequently. In fact, if we normalize the height of each bar in a histogram, dividing it by the total number of signal samples that was used to create it, the resulting graph gives us an empirical sense of "how likely it is that the random signal used to build the histogram will take a value within the range represented by a given bar." For example, if we use 1000 samples of a random signal to develop a histogram and 126 are placed in the histogram bin from 2.0 to 2.1, this will be telling us that 126 / 1000 = 0.126 or 12.6% of this segment of the random variable samples had an amplitude between 2.0 and 2.1. Now, if we were to take another 1000 points *of the same random variable* and develop the same kind of histogram we could provide an educated guess in answering the question: What is the probability that in this new segment a sample will land in the bin from 2.0 to 2.1? We could answer 12.6% (or 0.126). In other words, the normalized histogram provides us with an empirical (coarse) way to determine the probability that a certain random variable (e.g., the amplitude of the random signal) will land in a given interval.

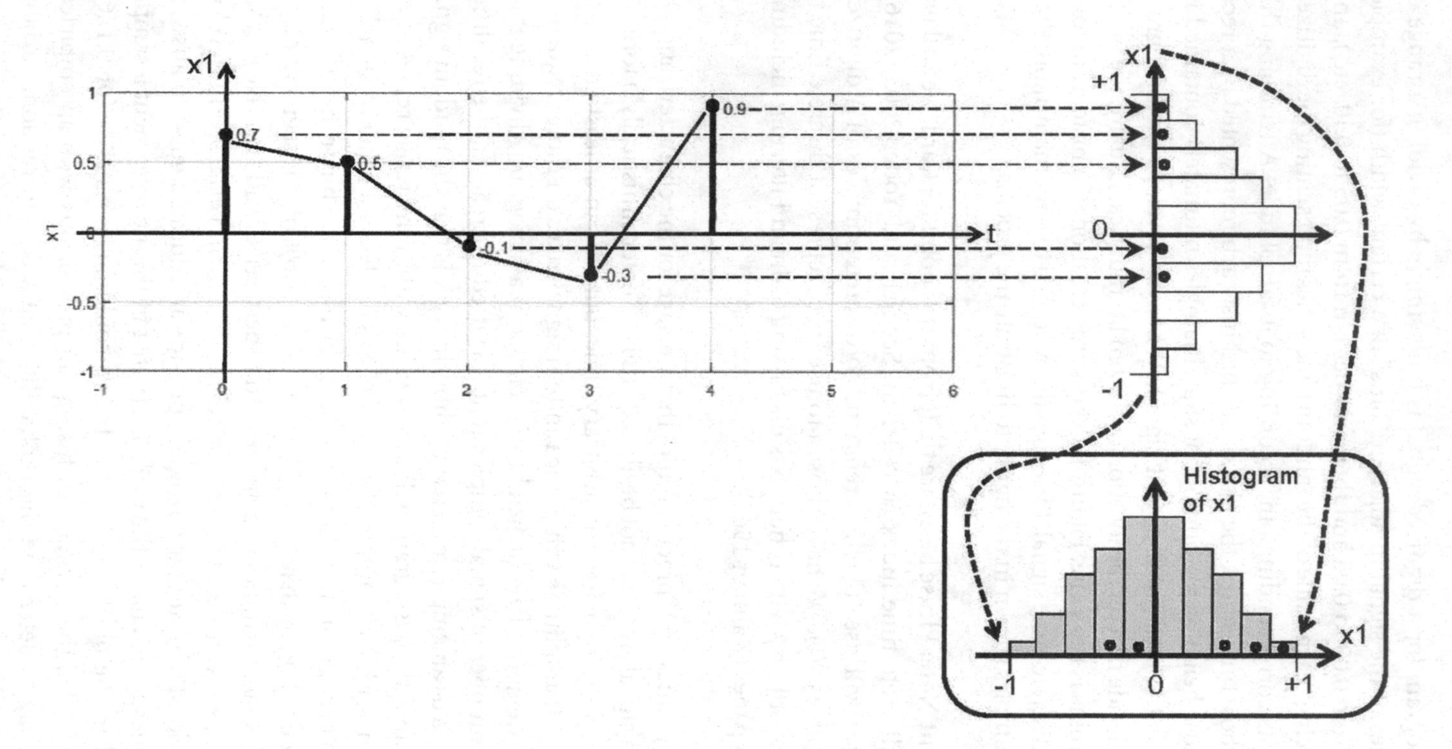

FIGURE 1.2 Creation of the histogram of a random discrete time signal. (Only five samples are shown, as examples.)

As an example, consider the creation of the following random discrete signal in MATLAB® and the development of its normalized histogram (also in MATLAB®). Please note that the MATLAB® command "rng(12345, 'v5normal')" resets the random number generator in MATLAB®, loading the seed value 12345. This will make it possible to obtain repeatable (pseudo) random sequences every time this script is executed. If that is not desired, the line with the rng command can simply be "commented out," by typing a "%" character at the beginning of that line.

```
%%% MATLAB CODE 01.01 ++++++++++++++++++++++++++++++++
rng(12345,'v5normal'); % Resets the Random Number
% Generator to ver5. Normal, seed = 12345;
N = 10000;  % Total number of signal samples
x = randn(N,1);
% NOTE: M' yields the conjugate transpose of matrix M.
% If M is real, M' just transposes it
n = linspace(0,(N-1),N)';
figure; plot(n,x,'k'); grid;
xlabel(' Discrete time index t');
ylabel('x(t)');

% For the Normalized Histogram
% [N x binwidth]/NormFactor = total_area/F = 1.0
% -> NormFactor = N x binwidth
binwidth = 0.1;
EDGES = [-3.0:binwidth:3];
NormFactor = N * binwidth;
figure; bar(EDGES,((histc(x,EDGES))/
NormFactor),0.95,'k');grid;
xlabel('x');
ylabel('Normalized histogram of x');
%%% MATLAB CODE 01.01 ++++++++++++++++++++++++++++++++
```

The normalized histogram shown in Figure 1.4 conveys some interesting information. It tells us that the values of the random digitized signal, or "time series" shown in Figure 1.3, occur with almost equal frequency above and below zero. It also tells us that very few, if any, samples had a value of more than 3 or less than -3. Further, there is a distinct pattern in which the "probability" that a sample value is contained in a given bin decreases from the bins closest to 0 to those close to 3 or -3. It is also clear that the bin that captured the largest number of samples (i.e., the

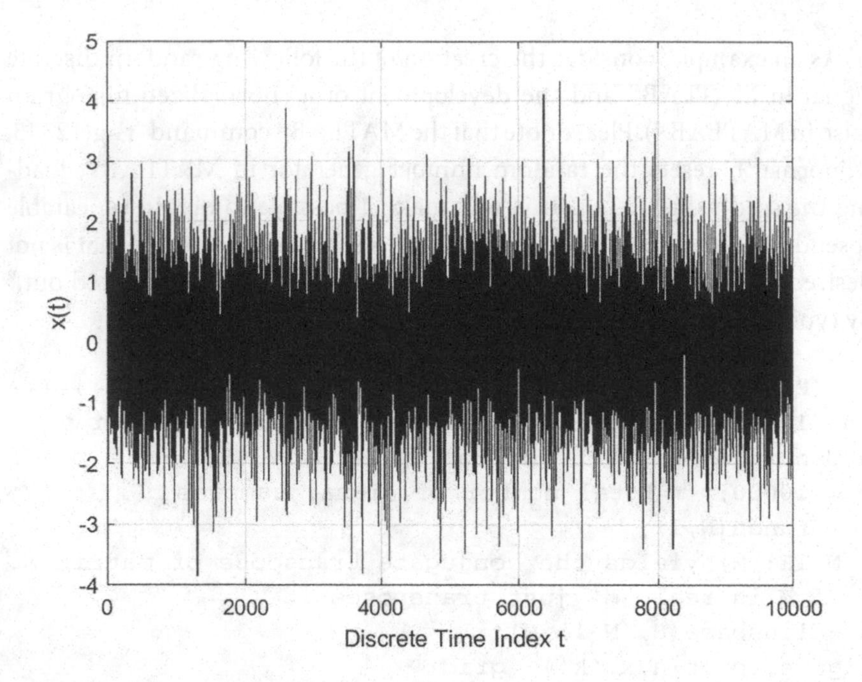

FIGURE 1.3 10,000 points of a random time series with mean 0 and variance 1.0.

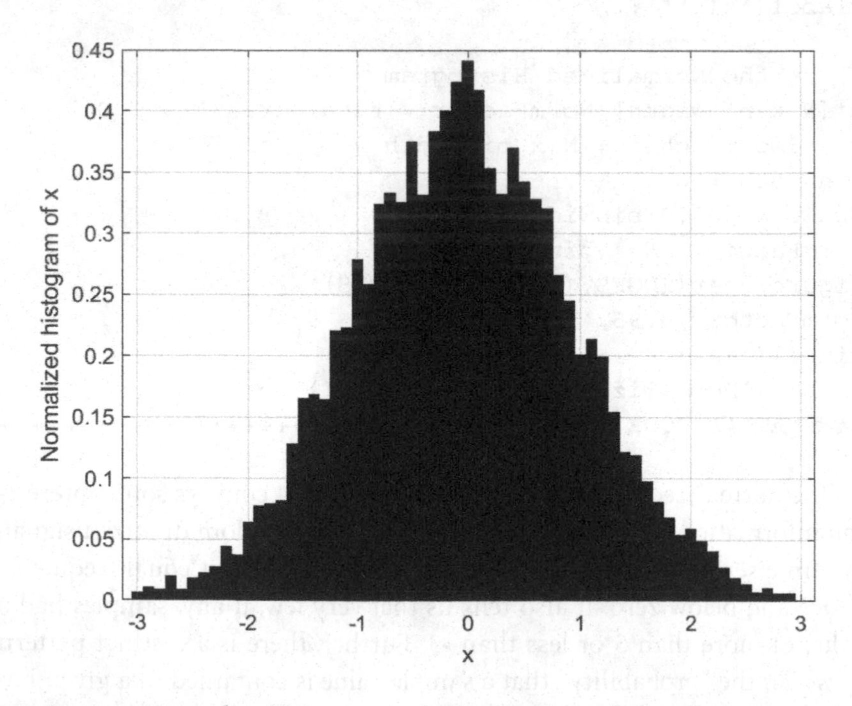

FIGURE 1.4 Normalized histogram of 10,000 points of a random time series with mean 0 and variance 1.0.

"most likely value of the samples") is, in this case, the bin that contains the value of zero. Because of the symmetry of the normalized histogram in Figure 1.4, we can derive that this bin with the highest bar in the normalized histogram is the value that we would obtain if we added all the values of the 10,000 samples processed to create the histogram and divided that sum by 10,000 (i.e., calculating the average). Of course all the values contained in the bin at zero would average to that value (0) and for a given amount of samples in the bar at -1 (for example) there seems to be an approximately equal number of samples captured by the bin at +1. Therefore, the contributions of these two symmetric bars (-1 and +1) to the overall average will not push it above or below the zero value. And this seems to be the case for any pair of bars placed symmetrically to the left and right of that most probable value (0).

A word should be said here about the way in which the normalization was performed in the previous MATLAB® script. Note that we incremented the number of samples for a given bin (for example the one between 2.0 and 2.1), for as long as the amplitude of the discrete-time sequence under consideration fell *anywhere* within that interval. When we observed a sample of amplitude 2.02, we incremented the count for that bin. If we observed a sample of amplitude 2.07, we incremented the count for that bin, etc. Therefore, the total count in a given bin will reflect the probability of finding a value of the random variable throughout the interval that spans what we have called "binwidth" (in the MATLAB® script earlier binwidth = 0.1). Graphically, the probability of that interval is represented as the rectangle for that bin, which has a shaded rectangular area of: base x height = binwidth x bin_count. So, the "total probability" for this random signal (if we placed all or nearly all the samples analyzed into one of the available bins) will be the sum of all the probabilities represented by shaded rectangles, i.e.,

$$\begin{aligned}(binwidth * bin\_count1) + \ldots + (binwidth * bin\_countLast) \\ = binwidth * TotalCount\end{aligned} \tag{1.13}$$

Again, if all (or nearly all, for an approximation) the samples of the random signal were placed in one of the available bins, the total probability (i.e., the probability that the random variable took on "any" value) must be 1.0 (i.e., certitude). Therefore, using N to represent the total number of samples processed for the histogram: binwidth x N should represent a probability of 1.0.

Since the counts for each of the bins are integer positive numbers (or 0), their total sum will be a positive number bigger than 1. Therefore, a normalizing factor (NormFactor) must be used to divide all the bin counts in such a way that, after that normalization, the sum of all the rectangular bar areas will be (approximately) 1. From the previous equation, the normalizing denominator needed is:

$$NormFactor = binwidth * N \tag{1.14}$$

This normalization is performed in the MATLAB® script and yields the normalized histogram in Figure 1.4.

1.3 THE GAUSSIAN (NORMAL) DISTRIBUTION

While I could now characterize the random signal for which I drew the normalized histogram by sending a picture of it or the locations and heights of all its bars to my foreign colleague, there is a much more compact mathematical model to encapsulate all this information. The shape of the normalized histogram in Figure 1.4 is found in many cases in nature (just like we find many circles in nature), and a generic mathematical description of this bell shape is available (Mix 1995; Papoulis and Pillai 2002; Yates and Goodman 2014; Hayes 1996):

$$pdf(x) = \frac{1}{\sqrt{2\sigma^2\pi}} e^{\frac{-(x-\mu)^2}{2\sigma^2}} \tag{1.15}$$

This equation yields the "probability density function," pdf, of our random variable, because it can be used to calculate the probability that the random variable x will be contained in a small interval of values of x (much like each one of the bins in the normalized histogram did). This probability density function, in particular, is called a Gaussian pdf, or Gaussian distribution. As mentioned before, this is such a common pdf that it is also known as a "Normal" pdf or a "Normal" distribution (Snider 2017). This is a continuous function of the random variable x. If you evaluate this formula for a large range of values of x, the values of pdf(x) obtained will always describe a bell shape. However, the position of its center and its width will vary, depending on the values of the parameters μ (mean) and σ^2 (variance) used. The mean determines the value of x where the center (axis of symmetry) of the bell shape will be. Therefore, $x = \mu$ will have the highest value of pdf(x). In other words, the most likely value of a random variable with this kind of probability density function is the mean, μ. The variance, σ^2, is associated with the "spread" of the different values taken by the random variable x above and below (i.e., higher than or lower than) the mean. Sometimes it is better to use the square root of the variance: σ, called the standard deviation, to characterize the width of the pdf. In any case, if the values of the random variable x are closely clustered around the mean, the bell shape will be "narrow" and the variance (or standard deviation) will be small in value. If the values of the random variable x are widely dispersed around the mean, the bell shape will be "wide" and the variance (or standard deviation) will be larger in value.

Whether the bell shape is narrow or wide, the area under the curve (from -infinity to +infinity) must always amount to 1.0. This is because of the same reason why the sum of the number of samples captured by all the bars in the histogram must amount to the total number of samples involved in drawing the histogram. (Another way to look at this is to consider that the probability that a sample of the signal could take on a value within the whole range, from -infinity to +infinity, is . . . well 100%!.)

Figure 1.5 shows the result (solid trace) of using Equation 1.15 to find pdf results for values of x from –5 to +5, when the parameters considered were $\mu = 0$ and $\sigma^2 = 1.0$. It should be noted that the profile of this "abstract" pdf model matches very closely the practically determined profile of our normalized histogram.

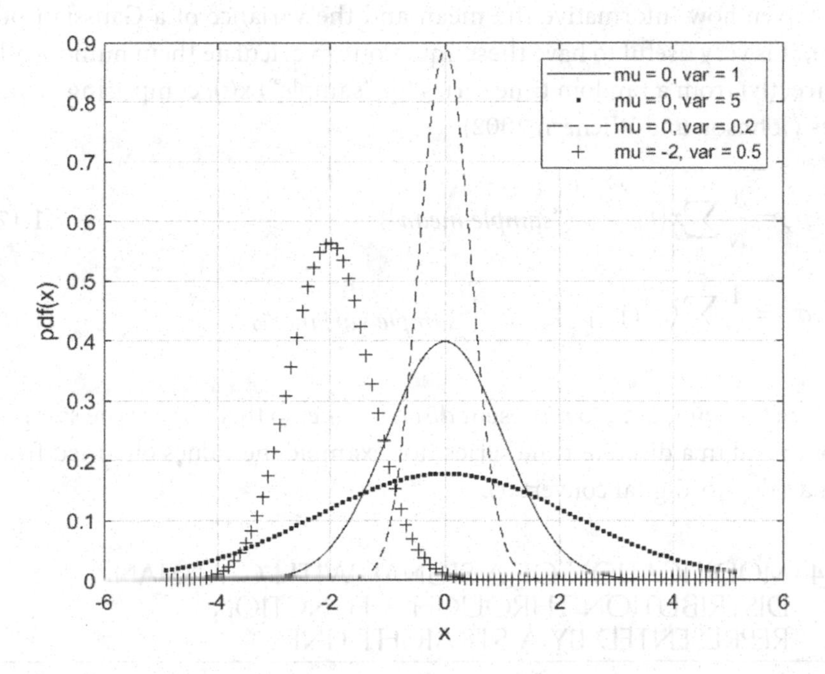

FIGURE 1.5 Gaussian Distributions. The solid trace corresponds to a Gaussian distribution with 0 mean and variance of 1. Three other combinations of mean and variance are also shown.

Because the Gaussian (Normal) distribution and its formula are well known everywhere, I could now characterize my random variable to my foreign friend by saying that I am modeling it as a Gaussian (Normal) probability density function with mean $\mu = 0$ and variance $\sigma^2 = 1.0$. Furthermore, also by convention, I could use well-established notation (McCool 2012) to just say that I am modeling the pdf of my random variable as:

$$\mathbf{N}\left(\mu,\sigma^2\right), \textit{which in this case would be } \mathbf{N}\left(0.0,1.0\right) \tag{1.16}$$

This is a highly compact form to characterize what we can actually say about the random signal that we are dealing with. With this piece of information, my friend could plug in $\mu = 0$ and $\sigma^2 = 1.0$ in Equation 1.15 and then draw the plot of the bell shape. With that she could estimate how likely it will be to observe the random variable take a value within any range (which would be the area under the pdf curve between the beginning and end values of the range in question). Additionally, the plot (or, for that matter the numerical values of μ and σ^2), will immediately reveal where is the center of the pdf, which is the most likely value of x to be observed, and how widely spread the occurrences will be around that center.

Given how informative the mean and the variance of a Gaussian pdf are, it is very useful to have these equations to calculate them numerically (directly) from a random time series (or "sample") x(t), comprising N values (Johnson and Wichern 2002):

$$\mu_x = \frac{1}{N}\sum_t x(t) \qquad (\textit{"sample mean"}) \tag{1.17}$$

$$\sigma_x^2 = \frac{1}{N}\sum_t \left(x(t) - \mu_x\right)^2 \qquad (\textit{"sample variance"}) \tag{1.18}$$

where t is being used to represent *discrete time*, so that x(t) are the samples contained in a discrete-time series (for example the values obtained from an analog-to-digital converter).

1.4 MODIFICATION OF A SIGNAL WITH GAUSSIAN DISTRIBUTION THROUGH A FUNCTION REPRESENTED BY A STRAIGHT LINE

Another important consideration for our study is to discuss what happens to a Gaussian random variable (in terms of its mean and its variance) when the samples of the original random time series x(t) are processed by a function represented by a straight line. This is the case of a transformation that just multiplies the input signal samples by a constant coefficient and adds a constant value to the product, following a formula such as this:

$$y(t) = mx(t) + b \tag{1.19}$$

A straight line, with slope m and intersecting the vertical axis at b, can be used to illustrate the projection of every sample of the input time series x(t) to become the corresponding output sample of the resulting time series y(t). So, for example, if a transformation is given by

$$y(t) = 2x(t) + 1.5 \tag{1.20}$$

Figure 1.6 shows how a few samples of the random signal x(t) used in the previous discussion would be projected to y(t).

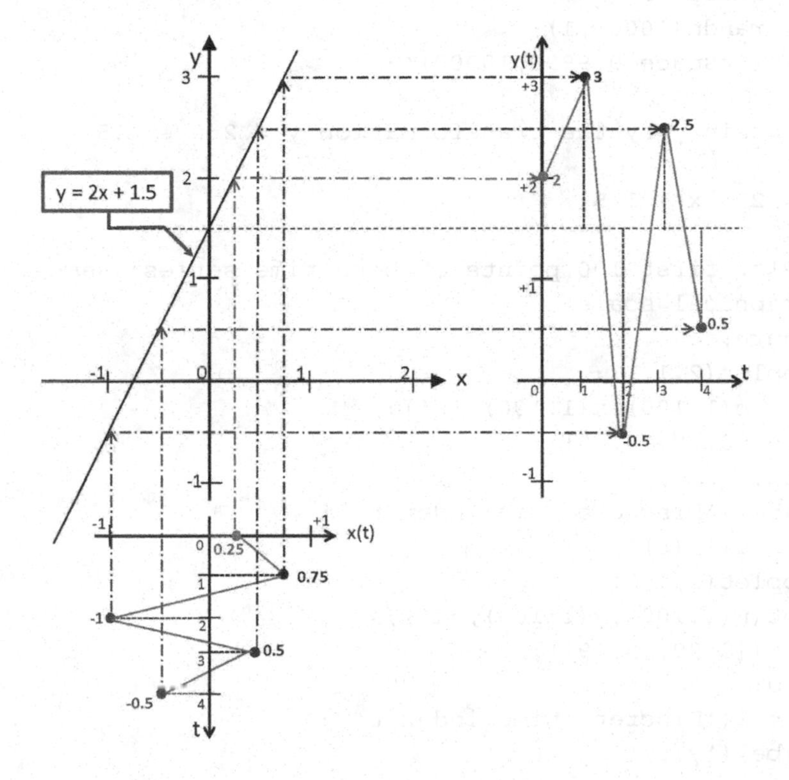

FIGURE 1.6 Projection of some samples of the input sequence x(t) to the output sequence y(t).

Note: A function such as the one indicated by Equation 1.20 cannot be called a "linear transformation" because it does not have the properties of homogeneity and additivity which are required to consider a transformation "linear" (Lathi 1998). For example, for x = 1, y = 3.5; for x = 4, y = 9.5; but, for x = 1 + 4 = 5, y will be 11.5, and *not* 3.5 + 9.5 = 13. The transformation indicated by Equation 1.19 is truly linear if b = 0.

If we project all 10,000 samples of x(t) to obtain the corresponding 10,000 samples of y(t), the comparison of the first 100 samples of both signals is shown in Figure 1.7. The comparison of their normalized histograms is shown in Figure 1.8. These figures were obtained with the following MATLAB® commands:

```
%%% MATLAB CODE 01.02 +++++++++++++++++++++++++++++++
rng(12345,'v5normal'); % Resets the Random Number
%Generator to ver5. Normal, seed = 12345;

% Generate x
x = randn(10000,1);
n = linspace(0,9999,10000)';

% Obtain y by the transformation y = 2 x + 1.5

y = 2 * x + 1.5;

% Plot first 100 points of both time series (same
%graphical scale)
figure;
subplot(2,1,1);
plot(n(1:100),x(1:100),'k');
axis([0,99,-9, 9]);
grid;
xlabel('Discrete time index t');
ylabel('x(t)')
subplot(2,1,2)
plot(n(1:100),y(1:100),'k');
axis([0,99,-9, 9]);
grid;
xlabel('Discrete time index t');
ylabel('y(t)')

% NORMALIZED HISTOGRAMS
N = 10000;
```

```
EDGES = [-9.0:0.1:9];
binwidth = 0.1;
NormFactor = N * binwidth ;

figure;
subplot(2,1,1);
bar(EDGES,((histc(x,EDGES))/NormFactor),1.02,'k');grid;
axis([-9,9,0, 0.5]);
xlabel('x');
ylabel('Normalized Histogram of x')

subplot(2,1,2);
bar(EDGES,((histc(y,EDGES))/NormFactor),1.02,'k');grid;
axis([-9,9,0, 0.5]);
xlabel('y');
ylabel('Normalized Histogram of y')
%%% MATLAB CODE 01.02 ++++++++++++++++++++++++++++++++
```

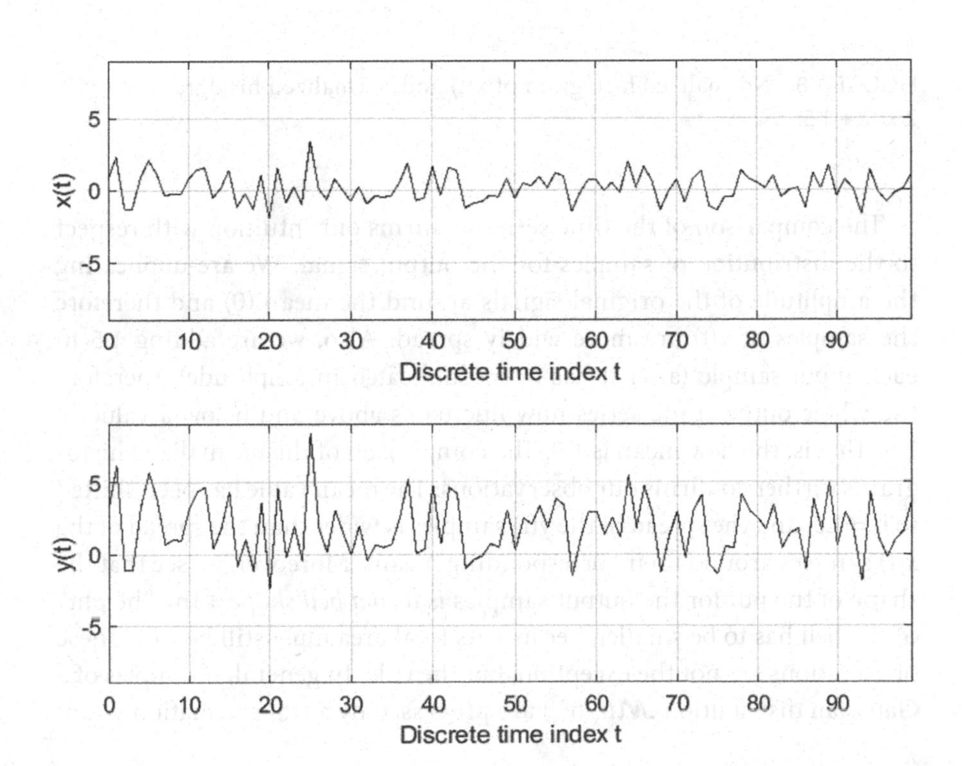

FIGURE 1.7 First 100 samples of x(t) and first 100 samples of y(t) = 2 x(t) + 1.5.

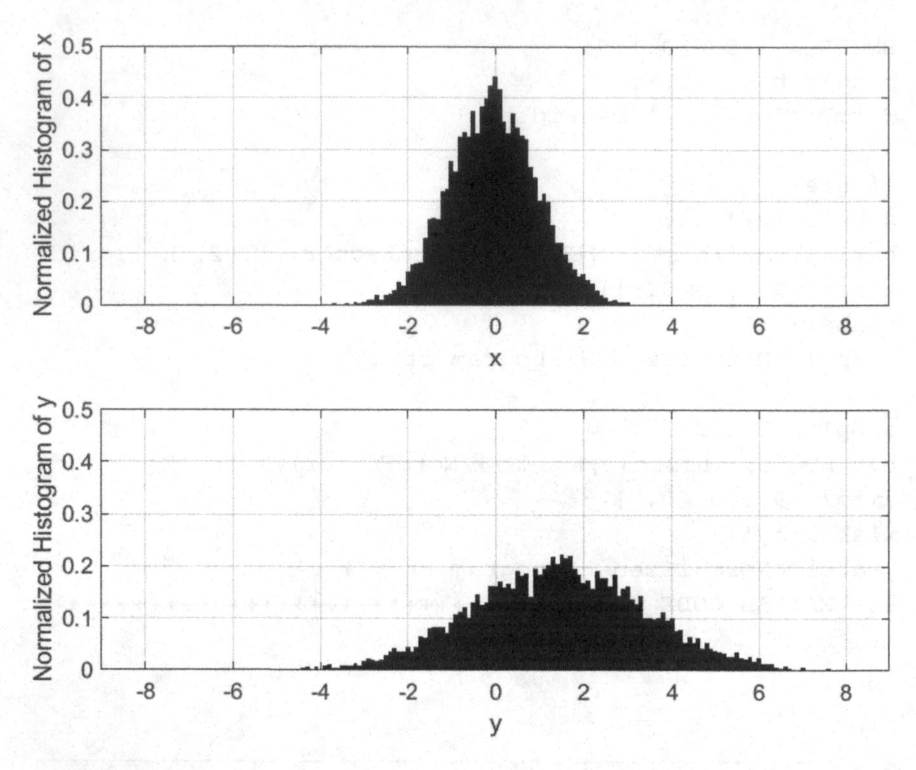

FIGURE 1.8 Normalized histogram of x(t) and normalized histogram of y(t) = 2 x(t) + 1.5.

The comparison of the time series confirms our intuition with respect to the distribution of samples for the output signal: We are duplicating the amplitude of the original signals around the mean (0) and therefore the samples of y(t) are more widely spread. Also, we are adding 1.5 to each input sample (after it has been duplicated in amplitude), therefore the whole output time series now fluctuates above and below a value of 1.5. That is, the new mean is 1.5. The comparison of the normalized histograms further confirms our observations: The mean value has been shifted to b = 1.5, and the spread of the y(t) samples is wider than the spread of the x(t) samples around their corresponding means. Moreover, we see that the shape of the pdf for the output samples *is also a bell shape.* (The "height" of the bell has to be smaller, because its total area must still be 1.0.) These observations are not the exception, but the rule. In general, if samples of a Gaussian distribution $\mathcal{N}(\mu_x,\sigma^2_x)$ are processed by a transformation y = m x + b, then the output random variable will have a Gaussian distribution with mean

$$\mu_y = m\mu_x + b \tag{1.21}$$

and variance

$$\sigma_y^2 = m^2\sigma_x^2 \tag{1.22}$$

In other words, the output of the transformation y = m x + b will have a Gaussian pdf:

$$\boldsymbol{N}\left(\mu_y, \sigma_y^2\right) = \boldsymbol{N}\left(m\mu_x + b, \sigma_x^2 * m^2\right) \tag{1.23}$$

Putting it into words, we would say that the mean of the resulting Gaussian is obtained by simply "processing" the mean of the input Gaussian. And the variance of the resulting Gaussian is obtained by multiplying the variance of the input Gaussian times m^2 (McCool 2012). It should be noticed then that the output Gaussian could be wider than the input Gaussian if m > 1 (as in our example) or it could be narrower, if m < 1. It is also apparent that the b constant in the transformation only has an impact on the mean of the output distribution, but does not affect the variance of the output distribution.

In the example

$$\mu_y = m\mu_x + b = (2)(0) + 1.5 = 1.5 \tag{1.24}$$

and

$$\sigma_y^2 = m^2\sigma_x^2 = \left(2^2\right)\left(1.0\right) = 4.0 \tag{1.25}$$

This is apparent in Figure 1.8 by the shifting and widening of the bell shape of the output histogram, in comparison to the bell shape of the input histogram.

One graphical way to represent these changes of the pdf through a transformation represented by a line is as in Figure 1.9.

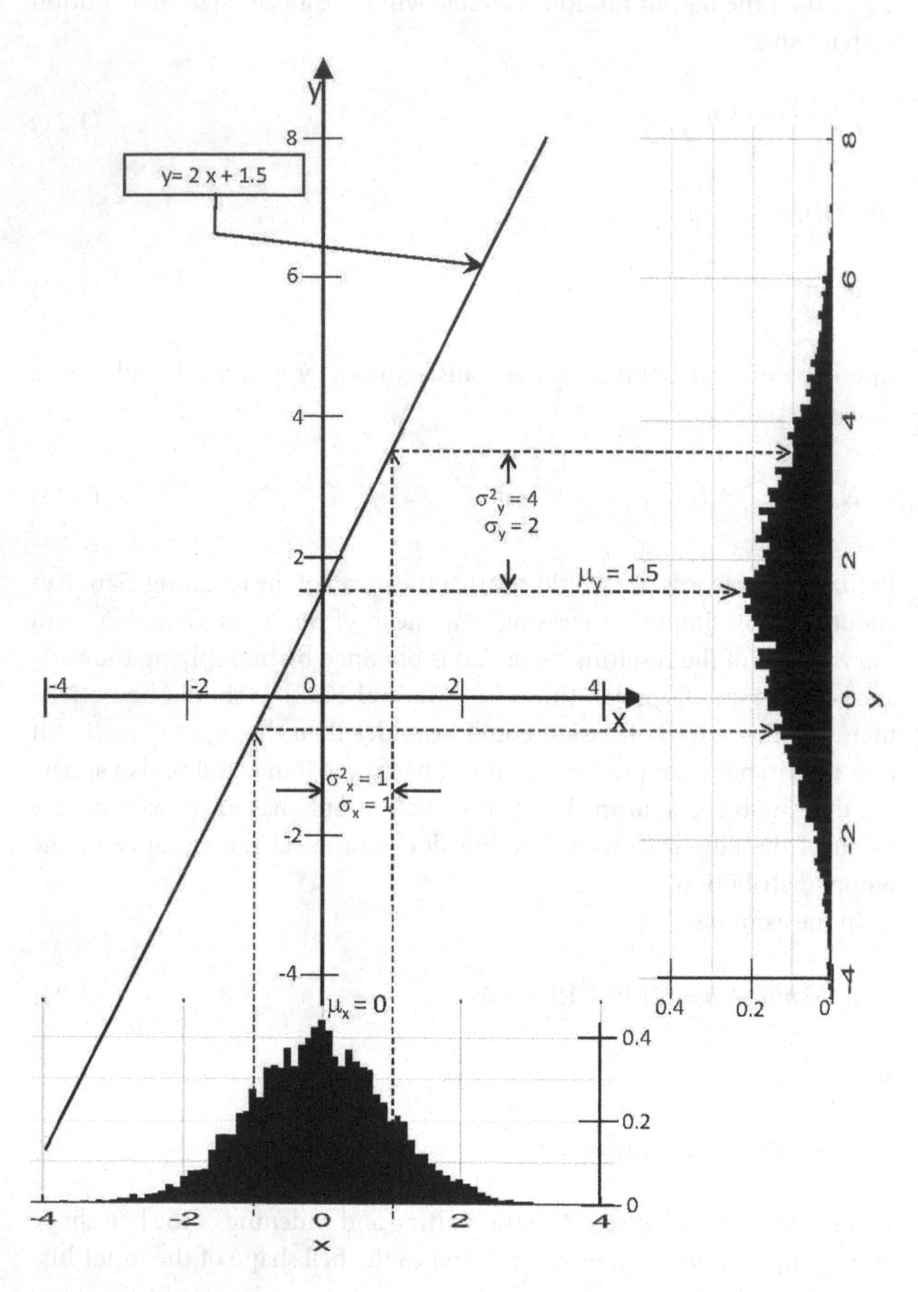

FIGURE 1.9 Effects of projecting a random variable, x(t) through the transformation y(t) = 2 x(t) + 1.5.

1.5 EFFECTS OF MULTIPLYING TWO GAUSSIAN DISTRIBUTIONS

Consider the product of two Gaussian distributions $\mathcal{N}(\mu_1, \sigma^2_1)$ and $\mathcal{N}(\mu_2, \sigma^2_2)$. Then, it is known that the mean and variance of the resulting Gaussian distribution $\mathcal{N}(\mu_p, \sigma^2_p)$ are (Bromiley 2003):

$$\mu_P = \frac{\mu_1\sigma_2^2 + \mu_2\sigma_1^2}{\sigma_1^2 + \sigma_2^2} = \mu_2 + \frac{\sigma_2^2(\mu_1 - \mu_2)}{\sigma_1^2 + \sigma_2^2} \tag{1.26}$$

and

$$\sigma_p^2 = \frac{\sigma_1^2\sigma_2^2}{\sigma_1^2 + \sigma_2^2} = \sigma_2^2 - \frac{\sigma_2^4}{\sigma_1^2 + \sigma_2^2} \tag{1.27}$$

(These two equations first present the most common formats for these results, and then they also present alternative formats, more appropriate for our discussions, which can be obtained after a few simple algebraic manipulations.)

And, if we define the parameter k as this ratio of variances:

$$k = \frac{\sigma_2^2}{\sigma_1^2 + \sigma_2^2} \tag{1.28}$$

Then

$$\mu_P = \mu_2 + k(\mu_1 - \mu_2) \tag{1.29}$$

and

$$\sigma_p^2 = \sigma_2^2 - k\sigma_2^2 = \sigma_2^2(1 - k) \tag{1.30}$$

As an example, consider the product of these two Gaussian distributions: $\mathcal{N}_1(1, 1)$ and $\mathcal{N}_2(6, 0.25)$. First, using the original equations (Equations 1.26 and 1.27):

$$\mu_P = \frac{\mu_1\sigma_2^2 + \mu_2\sigma_1^2}{\sigma_1^2 + \sigma_2^2} = \frac{(1)(0.25) + (6)(1)}{(1) + (0.25)} = 5 \tag{1.31}$$

$$\sigma_p^2 = \frac{\sigma_1^2\sigma_2^2}{\sigma_1^2 + \sigma_2^2} = \frac{(1)(0.25)}{(1) + (0.25)} = 0.2 \tag{1.32}$$

And then with the equations that use the parameter k:

$$k = \frac{(0.25)}{(1)+(0.25)} = 0.2 \tag{1.33}$$

$$\mu_p = 6 + (0.2)(1-6) = 5 \tag{1.34}$$

$$\sigma_p^2 = (0.25)(1-(0.2)) = 0.2 \tag{1.35}$$

So, we verify that the original formulas, as well as the modified formulas (with the parameter k) yield the same results for the mean and the variance of the product.

We would like to call the reader's attention to an important and interesting observation. The characteristics of the resulting (product) Gaussian are influenced more by the original Gaussian that had a smaller variance. In our example $\sigma^2_2 = 0.25$ is smaller than $\sigma^2_1 = 1$. Notice how, indeed, $\mu_p = 5$ ended up being much closer to $\mu_2 = 6$ than to $\mu_1 = 1$. Also, the resulting variance, $\sigma^2_p = 0.2$ is closer to $\sigma^2_2 = 0.25$ than it is to $\sigma^2_1 = 1$. It is interesting that, actually $\sigma^2_p < \sigma^2_2$. Figure 1.10 helps to visualize these observations.

While we have referred in this last section to the product of two "Gaussian distributions," it should be understood that the conclusions reached are applicable when we multiply samples of data series

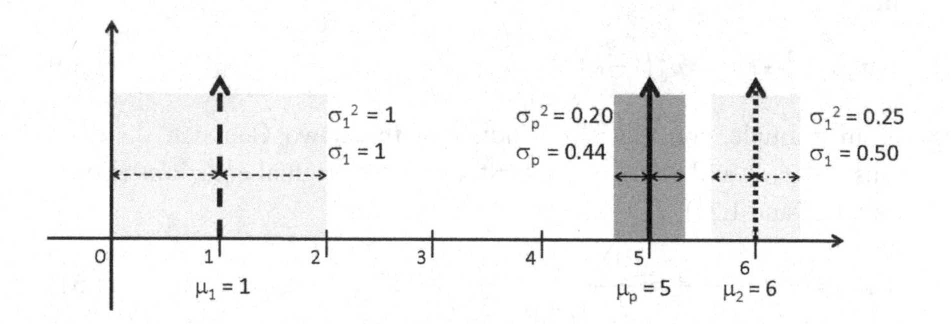

FIGURE 1.10 When a first Gaussian distribution (mean = 1, variance = 1) is multiplied with a second Gaussian distribution (mean = 6.0, variance = 0.25) the resulting Gaussian function has a mean (5.0) and a variance (0.2) that are closer to the mean and the variance of the original Gaussian with the lowest variance. The horizontal arrows span 1 standard deviation, in each case.

that have those Normal distributions. That is, if we multiply samples from a time series $x_1(t)$ whose probability function is $\mathcal{N}(\mu_1, \sigma^2_1)$ with samples of a second time series $x_2(t)$, whose probability function is $\mathcal{N}(\mu_2, \sigma^2_2)$, the resulting time series, $x_p(t)$, will have a Gaussian probability function $\mathcal{N}(\mu_p, \sigma^2_p)$, with μ_p and σ^2_p determined according to Equations 1.31 to 1.35.

CHAPTER 2

Multiple Random Sequences Considered Jointly

In our study of Kalman Filtering, it is most likely that we will not be handling a single random variable, in isolation, at a time. Instead, it is very likely that we will be dealing with circumstances where two or more random variables (possibly as two or more discrete-time series) will need to be considered jointly. The study of random variables provides a well-established framework for these cases and, fortunately, it is conceptually a natural extension of the ideas we have explored before, with some additional ideas that can only be applied when two or more random variables are being considered jointly. For the sake of simplicity, and also to have the opportunity to visualize the key concepts, the discussion here will be focused on cases when only two random variables are considered jointly. That is, we will consider "Joint distributions" of two random variables, which are referred to as "Bivariate distributions." However, the overall conclusions that we will reach are also applicable to the cases with more than two variables, such as the cases we will study in later chapters.

2.1 JOINT DISTRIBUTIONS—BIVARIATE CASE

So, let's consider that we have two discrete-time sequences:

$x_1(t)$

and

$x_2(t)$

Here, again, *t is being used as a discrete-time index*, so that in a practical scenario $x_1(t)$ and $x_2(t)$ could be series of numbers representing the samples digitized simultaneously by two analog-to-digital converters.

Each of these discrete-time sequences can be described by their own parameters:

$x_1(t)$ has mean μ_1 and variance $\sigma^2{}_1$
$x_2(t)$ has mean μ_2 and variance $\sigma^2{}_2$

In a practical scenario these parameters can be calculated by Equations 1.17 and 1.18 (from Chapter 1). Also, in a practical scenario we could draw histograms for each of the discrete-time series, separately.

However, when we consider them "jointly" (that is, as if we plotted both sequences according to the same discrete-time axis), additional aspects of their relationship can be characterized.

To follow the explanation, we need to shift our frame of mind somewhat. If we are considering them jointly, now we consider the value of x_1 at time t and the value of x_2 at the same time as *one* evaluation of this "joint distribution." Therefore, to obtain a "joint histogram" (2-D histogram) we need to place each joint value in a "square bin" within the 2-dimensional x_1, x_2 space. For example, now a bin may have the following four corners:

$x_1 = 0.0$ and $x_2 = 0.00$

$x_1 = 0.1$ and $x_2 = 0.00$

$x_1 = 0.1$ and $x_2 = 0.1$

$x_1 = 0.0$ and $x_2 = 0.1$

This is in contrast to the histogram bins for the univariate case for which we only needed to define "beginning" and "end" of the bin along a 1-dimensional scale.

Once it is finished and normalized (dividing the count of samples captured in each square bin by a normalizing factor that ensures that the total volume under the histogram adds to 1.0), a "2-D histogram" constructed in this way will give us an empirical sense of what is the probability that if I grab the pair of x_1 and x_2 values that occur at an arbitrary time t in these two sequences it will "fall" in a particular square bin, characterized by a range in x_1 and a range in x_2. Conceptually, this works just like it did when the 1-D histogram helped us to estimate the probability of a sample of x landing within the range of values of x associated with any specific 1-D bin.

Figure 2.1 illustrates how a few simultaneous samples of $x_1(t)$ and $x_2(t)$ would be placed in the bins of a 2-D histogram.

In the 1-D histogram, each bin resulted in a shaded rectangle and normalization was used to make the sum of all the shaded areas in the histogram add to 1.0. In the 2-D histogram, each bin will develop into a rectangular prism, containing a volume. In this case the normalization factor needed is such that the total volume of all the prisms will add to 1.0, after normalization.

As an example, consider the creation of two random discrete-time series, 100,000 points in length each, that occur simultaneously (i.e., for any discrete-time value t = 1 to 100,000, there is an $x_1(t)$ value and an $x_2(t)$ value). Since we will consider these two discrete-time sequences jointly, they will be created and kept in a single matrix, "dat," such that the whole column 1 of dat, i.e., dat(:,1) holds the sequence $x_1(t)$ and the whole column 2 of dat, i.e., dat(:,2) holds the sequence $x_2(t)$. In this scheme, the row numbers in dat coincide with the discrete-time index, t.

The MATLAB® script to follow creates the two random time sequences in the dat matrix and then they are used to create the 2-dimensional histogram of these sequences in square bins of 0.1 by 0.1 size from -2 to 6 in both axes.

The resulting (not normalized) histogram is displayed as a set of rectangular prisms (equivalent to the rectangular bars of the 1-D histogram). In Figure 2.2, this set of prisms is shown in a perspective view.

Then, the numerical values of the samples accumulated in each bin are stored in matrix N, which is transposed (to make the remaining plots match the first one in orientation) and normalized, dividing by the required normalizing factor, as described previously, to yield Matrix N1. Matrix N1 is used to generate alternative representations of the normalized 2-D histogram. Figure 2.3 displays it as a set of (colored) 3-D contours, showed in perspective view. It is important to notice how both the prisms and the 3-D contours represent the same shape (except for the scaling factor involved in normalization). As in the 1-D histogram case the highest point of the 2-D histogram represents the most frequently observed combination of x_1 and x_2 values, which is therefore the most probable one to occur. It should also be noted that the frequency of occurrences decreases gradually from that most frequent x_1, x_2 combination (same kind of decrease observed in the 1-D histograms we saw before). In MATLAB®, one could apply a 3-D rotation to view the 3-D contour plot from the top, or simply

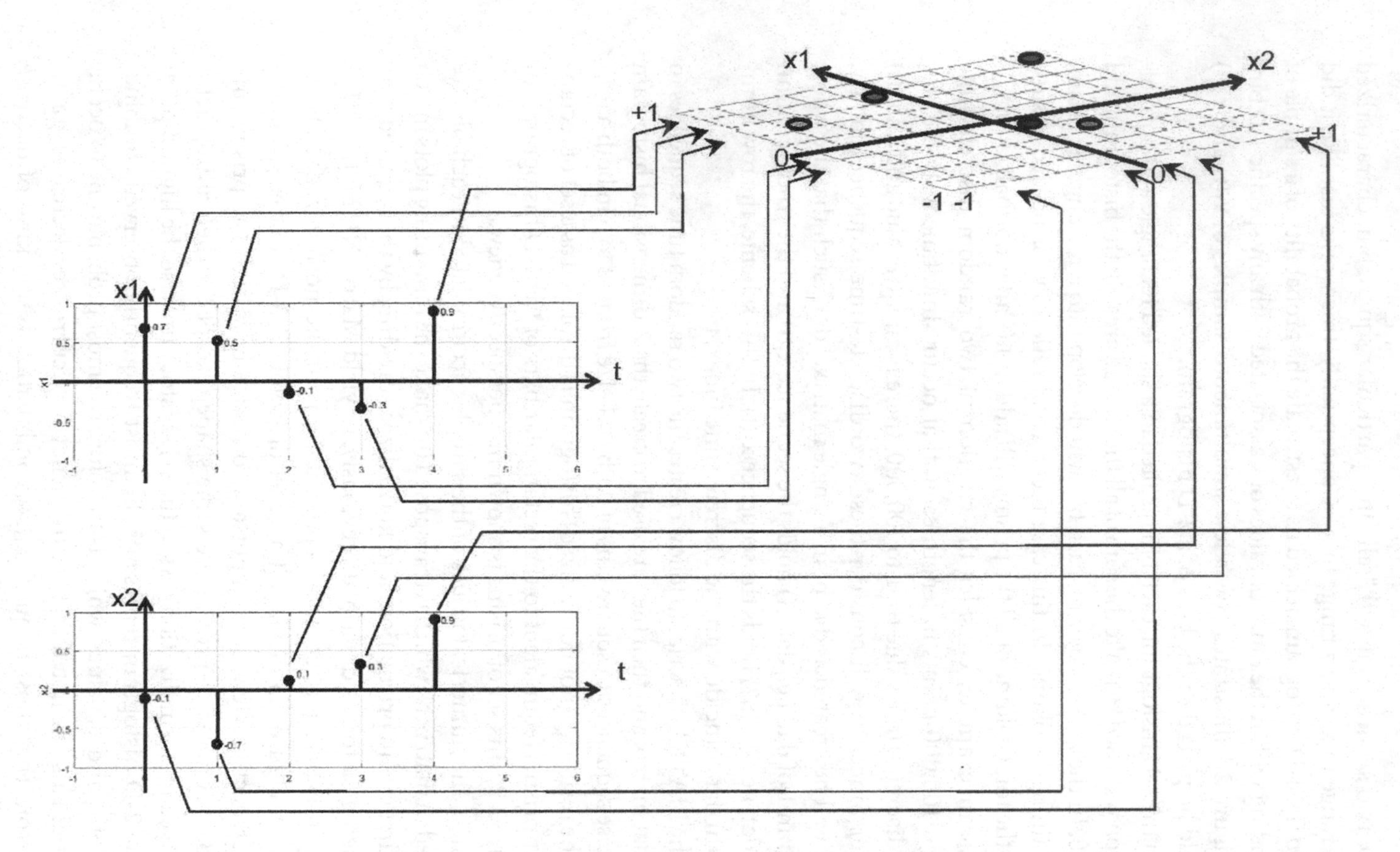

FIGURE 2.1 This figure provides a few examples of how pairs of $x_1(t)$ and $x_2(t)$ values, collected at the same time, would be placed in a 2-D histogram.

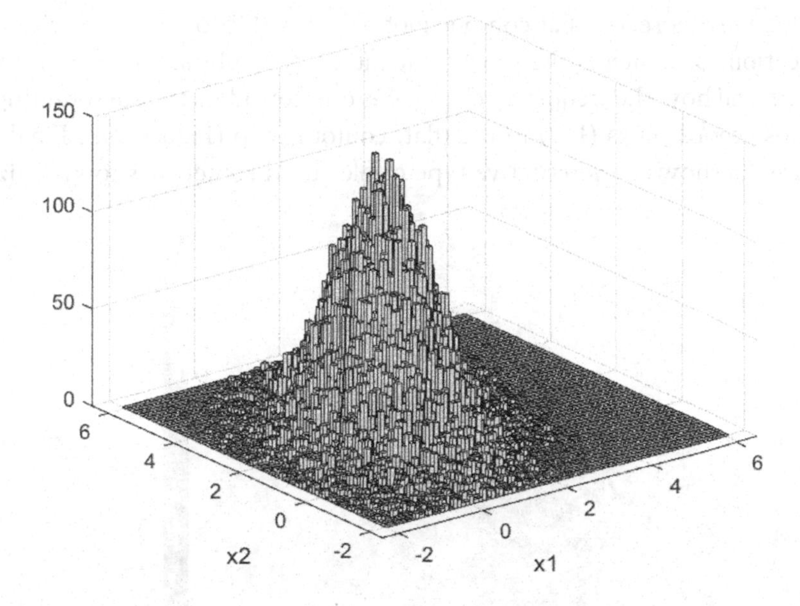

FIGURE 2.2 Perspective view of the 2-D histogram created from two discrete random sequences generated by the MATLAB® commands listed. Their means were 1 and 2, respectively. Their variances were 1 and 2, also. Their correlation was 0.5.

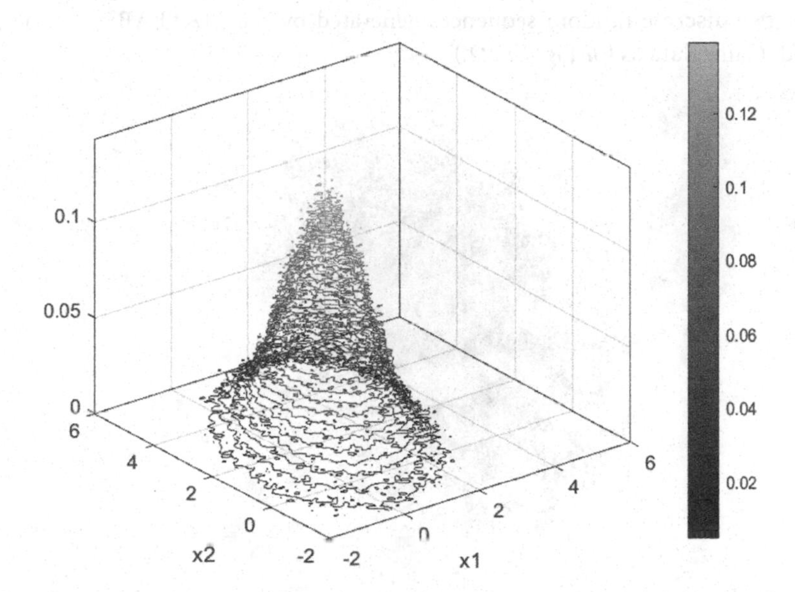

FIGURE 2.3 3-dimensional contour level plot of the 2-D histogram created from two discrete random sequences generated by the MATLAB® commands listed. (Same data as for Figure 2.2.)

(as done here) create a flat contour plot, which will provide much clearer indications of which is the most frequent x_1, x_2 combination ($x_1 = 1$ and $x_2 = 2$), and how the frequency changes as one considers the surroundings of those coordinates (1, 2) in the (flat) contour map (Figure 2.4). Finally, Figure 2.5 shows an alternative type of plot used sometimes to visualize

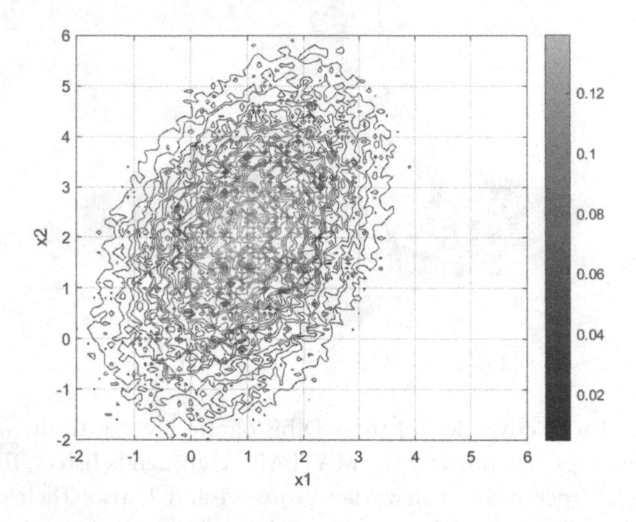

FIGURE 2.4 Top view of the contour level plot of the 2-D histogram created from two discrete random sequences generated by the MATLAB® commands listed. (Same data as for Figure 2.2.)

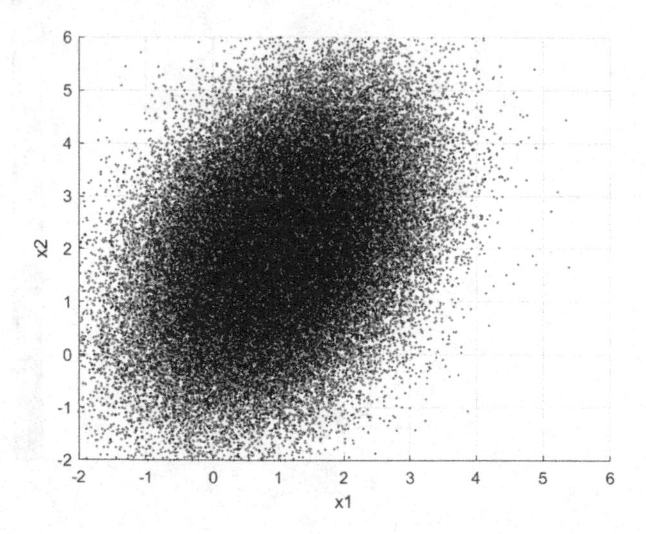

FIGURE 2.5 Density plot created from two discrete random sequences generated by the MATLAB® commands listed. (Same data as for Figure 2.2.)

similar information as contained in the 2-D histogram. That alternative display is sometimes called the "density plot." It is obtained directly from the raw values of both time series, using every $x_1(t)$ as abscissa and the corresponding $x_2(t)$ as the ordinate to plot one point. (This is achieved in MATLAB® with the "scatter" command.) Obviously, areas where x_1, x_2 combinations were more frequently found will end up with a higher density of the plotting color. These areas are located in the same regions where the tallest prisms were found within the original 2-D histogram.

```
%%% MATLAB CODE 02.01 ++++++++++++++++++++++++++++++++
% Creation of 2-time series with given means, variances
%and correlation
% Resetting RNG to ver5 Normal, seed = 12345.
% MEANS at mu and COVARIANCE MTX Sigma
% mu = [1,2]; Sigma = [1 .5; .5 2];

rng(12345,'v5normal');
mu = [1,2];Sigma = [1 .5; .5 2];
R=chol(Sigma);
Nsamp =100000; %Create Nsamp samples, each of the series
dat = repmat(mu,Nsamp,1) + randn(Nsamp,2) * R;
% The time series x1 & x2 are columns 1 & 2 of dat

% First Figure: Show PRISMS of 2-D histogram in
% perspective view
binside = 0.1;

figure;
x1edges = [-2.0:binside:6.0];
x2edges = [-2.0:binside:6.0];
edges{1} = x1edges;
edges{2} = x2edges;

    hist3(dat,'Edges',edges);%DRAW hist., perspective view
N = hist3(dat, 'Edges', edges ); %ON SAME DATA, get
%histogram values in matrix N for later use

    xlabel(' x1 '); ylabel(' x2 ');

% Display default 3D perspective view—NOT NORMALIZED:
view(3);
```

```
% NORMALIZATION and other forms of visualization
NormFactor = binside * binside * Nsamp;
N1 = N'/NormFactor;

% Second Figure: 3D contour levels in perspective
figure;
[n1r, n1c] = size(N1);

x1edges = [-2.0:0.1:6.0]; %These assignments are REPEATED
                           %here for clarity
x2edges = [-2.0:0.1:6.0]; %These assignments are REPEATED
                           %here for clarity


X1edgesm = repmat(x1edges, n1c, 1);
X2edgesm = repmat(x2edges',1,n1r);

contour3(X1edgesm,X2edgesm,N1,30);
colormap('winter');
grid on; xlabel('x1'); ylabel('x2'); view(3); colorbar;

% Third Figure; show the TOP VIEW (2D) of the contour
% plots for Normalized 2D Histogram

figure;
contour3(X1edgesm,X2edgesm,N1,30);
colormap('winter');
grid on; xlabel('x1'); ylabel('x2'); view(2); colorbar;

% Fourth figure 2D DENSITY PLOT

figure;
scatter( dat(:,1) , dat(:,2), 1 ) ;
grid on; xlabel(' x1 '); ylabel(' x2 '); view(2);
axis([ -2, 6 , -2 , 6])
%%% MATLAB CODE 02.01 ++++++++++++++++++++++++++++++++
```

2.2 BIVARIATE GAUSSIAN DISTRIBUTION—COVARIANCE AND CORRELATION

Just as the 1-D histogram was a practical (and imperfect) "incarnation" of the abstract model of a (1-D) Gaussian probability density function, the "mountain" shape of the 2-D histogram formed by the prisms seen in

Figure 2.2 is the practical portrayal of the "bivariate distribution" of these two time series. In this example, again, we are seeing a case of random variables that conform to the specific case of a "Gaussian" (or "Normal") bivariate distribution, for which the generic analytical expression has this format (Childers 1997; Stark and Woods 2002):

$$P(x_1, x_2) = \frac{1}{2\pi\sigma_1\sigma_2\sqrt{1-\rho^2}} exp\left[\frac{-z}{2(1-\rho^2)}\right] \tag{2.1}$$

where the term z is given by:

$$z = \frac{(x_1 - \mu_1)^2}{\sigma_1^2} - \frac{2\rho(x_1 - \mu_1)(x_2 - \mu_2)}{\sigma_1\sigma_2} + \frac{(x_2 - \mu_2)^2}{\sigma_2^2} \tag{2.2}$$

In Equation 2.2, the "ρ" is an encapsulated representation of a property of this bivariate distribution called the *correlation (coefficient)* between x_1 and x_2.

It may be preferable to address the meaning of the correlation by first discussing the closely related concept of *covariance.* The covariance is a similar concept to the variance that we have explained for the univariate case (Equation 1.18), but now it is defined in equal parts by the difference of values of the first variable (x_1) and its mean (μ_1) and by the difference of values of the second variable (x_2) and its mean (μ_2). For a specific pair of joint random sequences, $x_1(t)$ and $x_2(t)$, for which there are N pairs of values available, the (sample) covariance can be calculated as (Johnson and Wichern 2002):

$$COV(x_1, x_2) = \left(\frac{1}{N}\right)\sum_{t=1}^{N}(x_1(t) - \mu_1)(x_2(t) - \mu_2) \tag{2.3}$$

(As a mnemonic aid, notice that if we collapse the covariance of two variables to just one, making $x_2 = x_1$, then the result would collapse to the *variance* of that single variable.) But a very important difference in the calculation of the covariance is that both factors of each element of the summation appear as "first order terms." That is, the difference between a value of x_1 and its mean is not squared. The same occurs for the other factor within the summation: The difference between a value of x_2 and its mean is not squared. So, in this case of the covariance, some of the individual products that are being accumulated may be positive, but other individual products may be negative. (For the 1-D variance the terms within the summation are squared, hence they will all be positive contributions to the summation.) Consequently, we could have two random time sequences x_1 and x_2. both of which are widely spread about their corresponding means and still get a final value for the covariance calculation of 0 (or a small value)! But for that same scenario we could also end up with a very large final value of the covariance calculation. It all depends on whether or not the "signs" of the two $(x(t) - \mu)$ parentheses in each product are well coordinated, meaning "same sign," in which case the products will be predominantly positive, yielding a large positive value of covariance, or not. If the "signs" are not coordinated, meaning "opposite signs" in most cases, there will be a predominant number of negative products being added, which are likely to yield a large negative value as the result of the covariance calculation. While this precludes the use of the covariance to measure "spread," it brings up a different, but very important, interpretation of this attribute of two joint random discrete-time series. If we calculate a large positive covariance between x_1 and x_2 we will have reason to believe that for most values of t for which $x_1(t)$ takes on a value above its mean, we would observe that $x_2(t)$ also takes on a value above its mean (same positive sign for the differences $x(t) - \mu$). In that case (large positive covariance) when x_1 drops below its mean, x_2 is also likely to drop below its mean. In other words, a large positive covariance value indicates that these two discrete-time sequences increase and decrease (relative to their corresponding mean values) *simultaneously.* By a complementary argument, if the final result of the covariance calculation is a very large negative value we would know that most of the products in the summation that defines the covariance were negative. Therefore, the increase/decrease of one of the variables (relative to its mean) must have been predominantly "*opposite*" to the corresponding variations of the other variable about its mean. (In other words, when one went above its mean, the other one dropped below its mean.) Finally, if the covariance between two discrete-time random sequences $x_1(t)$ and $x_2(t)$ yields a value very close to zero (positive or negative) we would need to conclude that the two time series *do not seem to have much coordination* (direct or reversed) in how they vary about their means.

Going back to the "ρ" parameter in the definition of the 2-D Gaussian distribution (Equation 2.1), the correlation between x_1 and x_2, $\rho(x_1, x_2)$, is nothing more than the normalization of the covariance between x_1 and x_2, dividing $COV(x_1, x_2)$ by the product of the standard deviations (σ_1 and σ_2) of the time series:

$$\rho(x_1, x_2) = \frac{COV(x_1, x_2)}{\sigma_1 \sigma_2} \tag{2.4}$$

This normalization is actually very useful, because now we do not need to deal with ambiguous terms like "very large" positive covariance or "very large" negative covariance, since $\rho(x_1, x_2)$ will yield values only in the range −1 to 1 (Snider 2017). But, as explained previously, both the covariance and the correlation indicate the level of coordination in the variations of the two discrete-time random variables about their corresponding means.

Just as for the 1-D Gaussian case, if we started from the analytical expressions in Equations 2.1 and 2.2 and assigned specific values to define a (2-D) mean, for example $\mu_1 = 1$ and $\mu_2 = 2$, and values for the variances, for example $\sigma^2_1 = 1$ and $\sigma^2_2 = 2$, and a specific value for the covariance, such as $COV(x_1, x_2) = 0.5$, we could then compute numerical values for the 2-D Gaussian pdf of any x_1, x_2 combination. If we performed such evaluation at enough x_1, x_2 combinations and plotted the resulting pdf(x_1, x_2) values as the third ("vertical") dimension, visualizing the results in perspective view, we would verify that those pdf heights would define a surface that (approximately) envelopes our normalized 2-D histogram, as we saw previously for the 1-D pdf(x) plot and the normalized univariate histogram. An example of the perspective visualization of one such 2-D Gaussian distribution is shown in Figure 2.6. The mesh plot in this figure was created with the function msh2dg.m, listed in the following code.

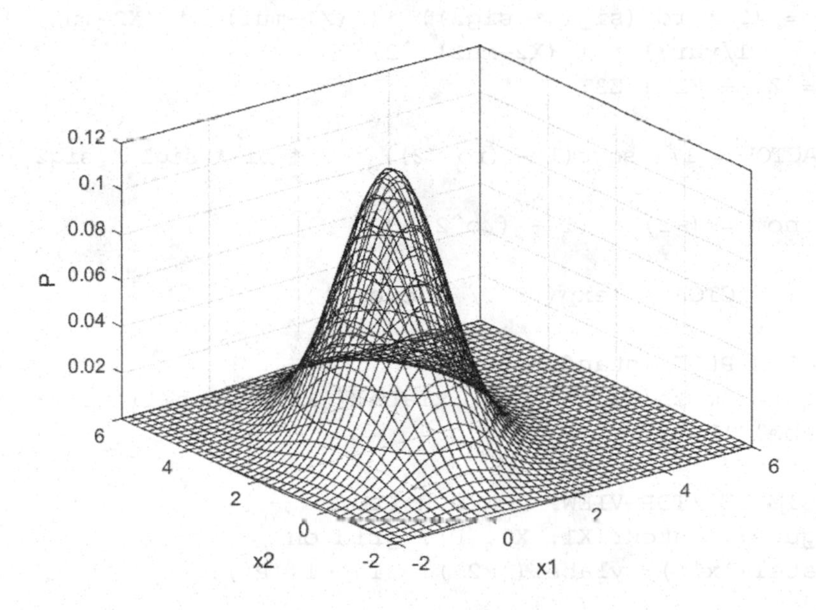

FIGURE 2.6 Mesh plot of a bivariate Gaussian distribution evaluated using the analytical expression in Equations 2.1 and 2.2. The parameters used were: Means: 1 and 2; Variances: 1 and 2; Covariance: 0.5.

```
%%% MATLAB CODE 02.02 ++++++++++++++++++++++++++++++++++
% msh2dg—creates & meshes a 2D Gaussian dist. x1, x2
%
% SYNTAX [X1, X2, P] = msh2dg(str, stp, nd, MU, SIGMA);
% MU is the vector of MEANS (e.g., [1;2])
% SIGMA is the COVARIANCE MATRIX (e.g., [1,0.5;0.5,2])
% (Covariance matrix is explained in Section 2.3)
% str(start),stp(step),nd (end) in MESHGRID(both x1 ,x2)
% Values used for example figure: -2, 0.2, 6
function [X1, X2, P] = msh2dg(str, stp, nd, MU, SIGMA);

var1 = SIGMA(1,1);
var2 = SIGMA(2,2);
sig1 = sqrt(var1);
sig2 = sqrt(var2);
ro = SIGMA(1,2)/(sig1 * sig2);
mu1 = MU(1);
mu2 = MU(2);

 [X1,X2] = meshgrid(str:stp:nd, str:stp:nd);

Z1 = (1/var1) * ( (X1—mu1).^2);
Z2 = (2 * ro/(sig1 * sig2)) *( (X1—mu1) .* (X2—mu2) );
Z3 = (1/var2) * ( (X2—mu2).^2);
Z = Z1 — Z2 + Z3;

PFACTOR = 1/( sqrt(1 — (ro ^2)) * 2 * pi * sig1 * sig2 );

edenom = (-2) * (1 — (ro^2));

P = PFACTOR * (exp( Z ./ edenom));

% MESH PLOT (standard)
figure; mesh(X1, X2, P);;xlabel('x1');ylabel('x2');
zlabel('P');

% SIMPLE (TOP VIEW) CONTOUR PLOT
figure; contour(X1, X2, P); grid on
xlabel('x1'); ylabel('x2'); zlabel('P');

% 3-D CONTOUR PLOT
figure; contour3(X1, X2, P); grid on
```

```
c = 0.8;
surface(X1,X2,P,'EdgeColor',[c,c,c],'FaceColor','none')
xlabel('x1'); ylabel('x2'); zlabel('P');

end % end of function msh2gd.m
%%% MATLAB CODE 02.02 ++++++++++++++++++++++++++++++++++
```

In both our normalized 2-D-histogram and the "theoretical" 2-D Gaussian pdf, we observe that the intersections defined by "cutting planes" parallel to the x_1, x_2 plane would be, generally, ellipses (Childers 1997). If COV(x_1, x_2) is zero and $\sigma_1^2 = \sigma_2^2$, then the Gaussian pdf will be symmetrical around the axis that is perpendicular to the x_1, x_2 plane and intersects it at the 2-D mean of the bivariate distribution, (μ_1, μ_2). In that case the elliptical contours described before would fall in the particular *circular* case of the ellipse format (Brown and Hwang 2012).

Drawing perspective views of these theoretical 2-D Gaussians or the associated 2-D normalized histograms is very laborious and they are not the most informative display type. Therefore, we may sometimes visualize Gaussian pdfs and normalized 2-D histograms through top-view contour level maps (Stengel 1994) such as the pair of images displayed in Figure 2.7. One of them is a repetition of the top-view contour map of the practical 2-D normalized histogram created with the MATLAB® script in Figure 2.7a. The accompanying image is a simple *sketch* using the elliptical

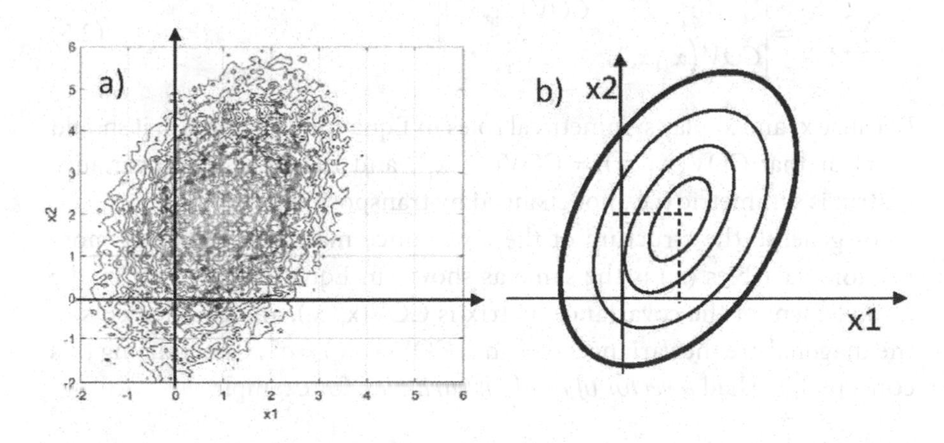

FIGURE 2.7 Comparison between a real 2-D histogram, represented by contour lines (a), and a simplified sketch indicating the (approximate) same characteristics of the bivariate distribution, such as means, variances and correlation (b).

contours (intersections between the 2-D Gaussian and horizontal planes) to provide a visualization of the corresponding 2-D pdf.

We should be able to use either of the representations compared in Figure 2.7 to gain an approximate sense for what is the 2-D mean of the bivariate distribution (μ_1, μ_2) and how widely spread the bivariate distribution is around that 2-D mean (area of the elliptical lowest contour of the distribution, plotted with a thicker line in Figure 2.7b).

2.3 COVARIANCE MATRIX

As we found for the case of a single random variable, we have now some alternatives for the characterization of a bivariate distribution. I could use a snapshot of the 2-D histogram; I could use the key parameters of the Gaussian bivariate distribution to plot/graph the abstract pdf(x_1, x_2) that I am proposing as the model for the bivariate distribution, or I could even just state that I am modeling the bivariate distribution as a 2-D Gaussian and list all the necessary parameters. If I decided to take the last approach, I could use a standard and highly compact way to describe the dispersion and coordination of the two variables. I could present the *covariance matrix* for these two joint random variables, $\Sigma_{x1,\, x2}$.

For two random variables, x_1 and x_2, considered jointly, the covariance matrix is given by:

$$\sum\nolimits_{x_1,x_2} = \begin{bmatrix} \sigma_1^2 & COV(x_1,x_2) \\ COV(x_2,x_1) & \sigma_2^2 \end{bmatrix} \tag{2.5}$$

Because x_1 and x_2 play symmetrical roles in Equations 2.3 and 2.4, it should be clear that COV(x_1, x_2) = COV(x_2, x_1), and therefore the covariance matrix is symmetric (i.e., not changed by transposition).

In general, the structure of the covariance matrix for three or more random variables (x_i) is the same as shown in Equation 2.5. That is, the (i, j) element of the covariance matrix is COV(x_i, x_j), and the elements in the diagonal are the variances, because COV(x_i, x_i) = σ^2_i. Generalizing this concept, if we had *a vector of random variables*, for example

$$\boldsymbol{x} = \begin{bmatrix} x_1 \\ x_2 \\ x_3 \end{bmatrix} \tag{2.6}$$

Then the covariance matrix for vector **x** would be:

$$\mathbf{\Sigma}_x = \begin{bmatrix} \sigma_1^2 & COV(x_1,x_2) & COV(x_1,x_3) \\ COV(x_2,x_1) & \sigma_2^2 & COV(x_2,x_3) \\ COV(x_3,x_1) & COV(x_3,x_2) & \sigma_3^2 \end{bmatrix} \quad (2.7)$$

which can also be obtained from the general definition of the covariance matrix for a *vector* **x** of d elements (Childers 1997):

$$\mathbf{\Sigma}_x = E\left\{(\boldsymbol{x}-\boldsymbol{\mu})(\boldsymbol{x}-\boldsymbol{\mu})^T\right\}$$

$$= E\left\{\begin{bmatrix} (x_1-\mu_1) \\ (x_1-\mu_1) \\ \vdots \\ (x_d-\mu_d) \end{bmatrix} \left[(x_1-\mu_1) \quad (x_1-\mu_1) \quad \cdots \quad (x_d-\mu_d)\right]\right\} \quad (2.8)$$

In this equation E{ } is the "expectation" operator, which can be seen, in this context, as obtaining an average, such that the value at the first row and first column of $\mathbf{\Sigma}_x$ is "the average" of $(x_1 - \mu_1)^2$, namely σ_1^2, as shown in Equation 2.7, which is the instantiation of Equation 2.8 for a vector **x** with only d = 3 elements.

2.4 PROCESSING A MULTIDIMENSIONAL GAUSSIAN DISTRIBUTION THROUGH A LINEAR TRANSFORMATION

In Section 1.4 we verified that, by applying a transformation such as y = mx + b to the samples of a single random sequence x(t) from a Gaussian distribution with mean μ_x and variance σ_x^2, we would be generating an output sequence y(t) whose values will have a Gaussian distribution with mean $\mu_y = m\,\mu_x + b$ and variance $\sigma_y^2 = (m^2)(\sigma_x^2)$. In a scenario in which we are handling multiple sequences, so that now **x** is not a scalar, but a vector, a linear transformation may generate multiple values in a vector **y** through a matrix product defined as:

$$\boldsymbol{y} = \boldsymbol{F}\boldsymbol{x} \quad (2.9)$$

$$\begin{bmatrix} y_1 \\ \vdots \\ y_N \end{bmatrix} = \begin{bmatrix} f_{11} & \cdots & f_{1M} \\ \vdots & \ddots & \vdots \\ f_{N1} & \cdots & f_{NM} \end{bmatrix} \begin{bmatrix} x_1 \\ \vdots \\ x_M \end{bmatrix} \quad (2.10)$$

If $\mathbf{x}$ is a vector of Gaussian random variables with mean vector $\boldsymbol{\mu}_x$ and covariance matrix $\boldsymbol{\Sigma}_x$, then the elements of the output vector also form a multidimensional Gaussian distribution (Johnson and Wichern 2002) (Peebles 2001) with its mean vector given by:

$$\boldsymbol{\mu}_y = \boldsymbol{F}\boldsymbol{\mu}_x \tag{2.11}$$

and its covariance matrix is given by:

$$\boldsymbol{\Sigma}_y = \boldsymbol{F}\boldsymbol{\Sigma}_x\boldsymbol{F}^T \tag{2.12}$$

where the "T" super-index indicates the transposition of the matrix.

2.5 MULTIPLYING TWO MULTIVARIATE GAUSSIAN DISTRIBUTIONS

Just as we presented in Chapter 1 formulas for the characteristics of the univariate Gaussian function that results when we multiply two univariate Gaussians, we will present here formulas for the characterization of the multidimensional Gaussian that results by multiplying two multidimensional Gaussians. In this case the two multidimensional Gaussians involved in the product are distributions $\mathcal{N}_0(\boldsymbol{\mu}_0, \boldsymbol{\Sigma}_0)$ and $\mathcal{N}_1(\boldsymbol{\mu}_1, \boldsymbol{\Sigma}_1)$, where now $\boldsymbol{\mu}_0$ and $\boldsymbol{\mu}_1$ are the vectors (for example d-by-1 vectors) containing the mean values of the multivariate distribution, and $\boldsymbol{\Sigma}_0$ and $\boldsymbol{\Sigma}_1$ are the corresponding covariance matrices (for example d-by-d matrices).

The literature (Ahrendt 2005; Gales and Airey 2006; Bromiley 2003) indicates that the product of those two Gaussian distributions will yield a Gaussian multidimensional function, such that:

$$\boldsymbol{N}(\boldsymbol{\mu}_0, \boldsymbol{\Sigma}_0)\boldsymbol{N}(\boldsymbol{\mu}_1, \boldsymbol{\Sigma}_1) = z_p\boldsymbol{N}(\boldsymbol{\mu}_p, \boldsymbol{\Sigma}_p) \tag{2.13}$$

where the covariance matrix of the product is:

$$\boldsymbol{\Sigma}_p = \left(\boldsymbol{\Sigma}_0^{-1} + \boldsymbol{\Sigma}_1^{-1}\right)^{-1} \tag{2.14}$$

And the mean vector of the product is:

$$\boldsymbol{\mu}_p = \boldsymbol{\Sigma}_p\left(\boldsymbol{\Sigma}_0^{-1}\boldsymbol{\mu}_0 + \boldsymbol{\Sigma}_1^{-1}\boldsymbol{\mu}_1\right) \tag{2.15}$$

Please note that the result involves a scaling factor z, which is (Ahrendt 2005):

$$z_p = \left|2\pi\left(\Sigma_0 + \Sigma_1\right)\right|^{-1/2} exp\left\{\frac{-1}{2}\left(\mu_0 - \mu_1\right)^T \left(\Sigma_0 - \Sigma_1\right)^{-1}\left(\mu_0 - \mu_1\right)\right\} \quad (2.16)$$

Our attention, however, will focus mainly on the mean vector and the covariance matrix of the multidimensional Gaussian function that results (without focusing much on the scaling factor). Fortunately, as it was the case in the univariate Gaussian product case (Section 1.5), there are some simpler and more indicative equations that can be used to obtain the mean vector, $\boldsymbol{\mu}_p$, and the covariance matrix $\boldsymbol{\Sigma}_p$, of the product:

$$\mu_p = \mu_0 + K\left(\mu_1 - \mu_0\right) \quad (2.17)$$

and

$$\Sigma_p = \Sigma_0 - K\Sigma_0 \quad (2.18)$$

where **K** is now a matrix:

$$K = \Sigma_0\left(\Sigma_0 + \Sigma_1\right)^{-1} \quad (2.19)$$

In this case, we will verify that Equations 2.14–2.15 and Equations 2.17–2.19 yield the same results in a specific example. Assume that the 2-by-1 mean vectors and 2-by-2 covariance matrices for two bivariate Gaussian distributions are given by:

$$\mu_1 = \begin{bmatrix} 1 \\ 1 \end{bmatrix} \;;\; \Sigma_1 = \begin{bmatrix} 1 & 0.25 \\ 0.25 & 1 \end{bmatrix} \quad (2.20)$$

$$\mu_2 = \begin{bmatrix} 6 \\ 6 \end{bmatrix} \;;\; \Sigma_2 = \begin{bmatrix} 0.25 & -0.025 \\ -0.025 & 0.25 \end{bmatrix} \quad (2.21)$$

This time we will make the calculations with the following MATLAB® script. We designate the results from Equations 2.14 and 2.15 with an "A" at the end of the variable names for the results (MUPA and SIGPA). The variable names for results obtained through Equations 2.17, 2.18 and 2.19 have a "B" at the end (MUPB and SIGPB). Note that this script calls the

MATLAB® function msh2dg, listed previously in this chapter (MATLAB® CODE 02.02).

```
%%% MATLAB CODE 02.03 ++++++++++++++++++++++++++++++++
% MULTIPLICATION OF 2 BIVARIATE GAUSSIAN DISTRIBUTIONS
MU0 = [1;1];
SIG0 = [1, 0.25; 0.25 , 1];
MU1 = [6;6];
SIG1 = [0.25, -0.025; -0.025, 0.25];
%%
%% Visualizing the first 2-D Gaussian: MU0, SIG0
[X01, X02, P0] = msh2dg(-2, 0.2, 8,MU0,SIG0);
%% Visualizing the second 2-D Gaussian: MU1 , SIG1
[X11, X12, P1] = msh2dg(-2, 0.2, 8, MU1, SIG1);
%
%% USING FORMULAS FROM Ahrendt, Bromiley ("FORMULAS A")
% Covariance matrix of the product is:
SIGPA = inv( (inv(SIG0)) + (inv(SIG1)) )
% Mean Vector of the Product Is:
MUPA =SIGPA * (((inv(SIG0)) * MU0)+((inv(SIG1)) * MU1))
%
% % NOW USING THE FORMULAS WITH K ("FORMULAS B")
%% The K matrix is:
K = SIG0 * ( inv( SIG0 + SIG1) )
% Covariance matrix of the product is:
SIGPB = SIG0 – ( K * SIG0)
% Mean Vector of the Product Is:
MUPB = MU0 + ( K * (MU1 – MU0))
%%
%% WE CONFIRM MUPA = MUPB and SIGPA = SIGPB
%%
%% Visualizing the product:
PP = P0.* P1; % Calculates the numerical point-to-point
% products of the distributions
figure;mesh(X11,X12,PP);
figure; contour(X11,X12,PP);grid
xlabel('x1');ylabel('x2');
%%% MATLAB CODE 02.03 ++++++++++++++++++++++++++++++++
```

The MATLAB® results show that, through both methods:

$$\mu_p = \begin{bmatrix} 5.2373 \\ 5.2373 \end{bmatrix} \quad ; \quad \Sigma_p = \begin{bmatrix} 0.1959 & -0.0053 \\ -0.0053 & 0.1959 \end{bmatrix} \tag{2.22}$$

This verified that both sets of equations provided the same results for the vector of means and covariance matrix of the product. Further, executing the MATLAB® code will create a number of figures for visualization of the three Gaussian distributions. Figure 2.8 shows only the contour level plots,

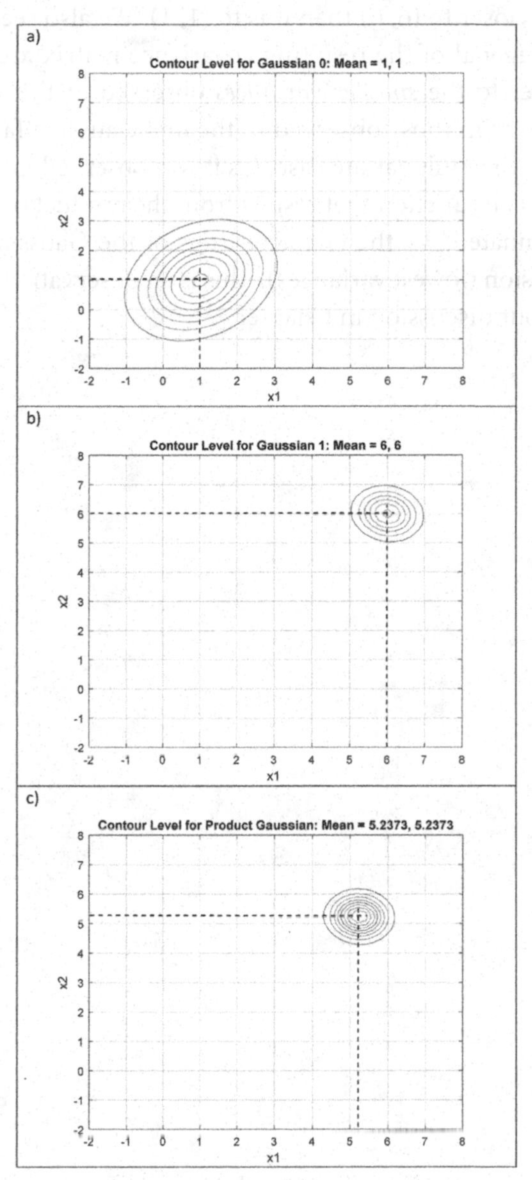

FIGURE 2.8 Contour level visualization of the first (Panel a) and second (Panel b) 2-D Gaussians being multiplied. Panel c shows the contour levels for the product Gaussian, whose mean vector and variances are more closely related to those of the original Gaussian *with the smallest variances.*

to emphasize how, as was the case for the product of univariate Gaussians (Section 1.5), the Gaussian with the lowest variances (the one with mean at 6, 6) has the largest influence in defining the characteristics of the product. We see in Figure 2.8c that the product Gaussian has its mean at (5.23, 5.23), which is much closer to (6, 6) than it is to (1, 1). We also see that the variances in the diagonal of the resulting covariance matrix are both 0.1959, which are closer to the *smaller variances* observed in the *second Gaussian factor* (0.25) than those observed in the first Gaussian factor (1.0). We observe that, in the multivariate case, (as it was observed in the univariate case), the Gaussian function that results from the product of two Gaussian factors is "dominated" by the characteristics of the Gaussian factor with the least dispersion (lowest variances). This is an observation which will be important for our discussion in Chapter 7.

CHAPTER 3

Conditional Probability, Bayes' Rule and Bayesian Estimation

In this chapter we introduce the reader to the concept of Bayesian Estimation, through which an *enhanced* estimate of a probability distribution, called the *Posterior Probability Distribution*, will be obtained by "merging" an initial *Prior Probability Distribution* with additional information, often from data observed or collected empirically from the specific situation being considered, encoded in a so-called *Likelihood Function*. The process of "merging" these two sources of information is termed Bayesian Estimation, as it evolves from the Bayes' Rule, which establishes an important relationship between two random variables and their conditional probabilities. A clear understanding of the concepts stated at the end of this chapter will be key during the discussion that will allow us to arrive to the Kalman Filtering algorithm, in Chapter 6.

3.1 CONDITIONAL PROBABILITY AND THE BAYES' RULE

To round up the background review on probability concepts that will be used in understanding the Kalman Filtering algorithm, we will now focus on how the probability of a random variable is different when no preliminary information is available, versus the case when some actual knowledge of facts that affect the random variable is available. We experience this kind of situation in many aspects of daily life. For example, sometimes we know that "if our favorite sports team manages to defeat their nemesis early in the playoffs" they will most likely end up taking the championship. Before our team plays against that tough early opponent we cannot say that it is very probable that they will win the championship overall. But, if the date of the tough playoff match passes and our team was victorious, we will probably begin to boast that our team has the championship "in the bag." That is, we will be estimating that the probability of our team becoming champions is very high.

If we say, for an example, that our team is TEAM_A and the tough opponent referenced in the previous paragraph is TEAM_B, another way to describe the scenario is to say that "the probability of TEAM_A becoming champions, *given that* TEAM_B *is defeated in the playoffs*, is very high." This is the typical format of a *conditional probability statement*, where whatever we may say about the probability being estimated (TEAM_A becoming champions) is being said *on the condition* that the precedent described (TEAM_B is defeated in the playoffs) is true. Frequently a vertical bar "|" is the syntax used to indicate the condition attached to this probability statement. So, for example, we could say:

$$p(TEAM\_A = champions \,|\, TEAM\_B = defeated\ in\ playoffs) = 0.95 \quad (3.1)$$

One key theorem (or "rule") related to conditional probabilities is "Bayes' Theorem" or "Bayes' Rule," attributed originally to Thomas Bayes, a XVIII-century English philosopher, preacher and mathematician.

Bayes' Rule states that (Stone 2015):

$$p(\theta|x)\,p(x) = p(x|\theta)\,p(\theta) \quad (3.2)$$

where:
$p(x)$ is the probability of x occurring.
$p(\theta)$ is the probability of θ occurring.
$p(x|\theta)$ is the (conditional) probability of x occurring once θ has occurred ("... given θ").
$p(\theta|x)$ is the (conditional) probability of θ occurring once x has occurred ("... given x").

While this "symmetric" formulation (Equation 3.2) may be easier to remember, an alternative presentation of Bayes' Rule reveals more directly the value that it has towards making "inferences" (Bayesian Inferences), through the combination of two probability assessments:

$$p(\theta \mid x) = \frac{p(x|\theta)\,p(\theta)}{p(x)} \tag{3.3}$$

This new presentation is particularly useful when we want to combine estimations of the two probabilities that appear in the numerator of the right-hand side of the equation to obtain the probability that appears on the left-hand term of the equation.

This idea of "combining" or "leveraging" two sources of information to yield a (more comprehensive) estimate of the probability on the left-hand side of the equation will be at the core of the Kalman Filtering concept.

For an initial understanding of the meaning and application of Bayes' Rule, as formulated in Equation 3.3, it is useful to refer to one example that appears in a number of books and has been popularized by Dr. James V. Stone, of the University of Sheffield, in his books (Stone 2013, 2015, 2016a, 2016b) and online contributions (Stone 2019).

The example focuses on determining (and comparing) the probability that a patient has *chickenpox*, (common and not usually deadly) *or smallpox* (at present very rare, but potentially deadly), given that the patient has spots on his skin.

We will use the values $\theta = 1$, or θ_1, for chickenpox and $\theta = 2$, or θ_2, for smallpox, and x will be used to indicate the presence of spots on the skin.

This example conveys to the reader the importance of "enhancing" our assessment of a situation by combining two sources of knowledge. It starts off by presenting an initial (and alarming) piece of knowledge: Imagine that a patient sees spots on his skin and goes to see a doctor. Now imagine that the doctor tells the patient: "It is known that 90% of patients with the dreaded smallpox will show spots on the skin as a symptom." So, the patient will be (justifiably) alarmed. But we would need to be careful to properly interpret this piece of information. How do we translate what the doctor told the patient to a conditional probability? Re-reading the doctor's statement we should realize that what he said is equivalent to:

$$p(x|\theta_2) = 0.9 \tag{3.4}$$

That is, most (90%) of patients with smallpox show skin spots. *But*, there are *other* possible reasons why a person may show skin spots. For example, it is also known that 80% of patients with the much less dangerous chickenpox display skin spots:

$$p(x|\theta_1) = 0.8 \tag{3.5}$$

Finding out about this "alternative" potential cause for his skin spots will probably calm the patient to some extent. Both $p(x|\theta_1)$ and $p(x|\theta_2)$ could be substituted as the first factor of the numerator in Bayes' Rule (Equation 3.3). Either of them can play the role called the "*Likelihood*" in the process of Bayesian Inference. It should also be noticed that this value is often determined by data observed in the specific situation at hand (i.e., the doctor *observed* that the patient has spots on his skin).

Because both likelihoods, $p(x|\theta_1)$ and $p(x|\theta_2)$ are close in value, one could have a first impulse to believe that it might be about equally possible for a patient to be affected by smallpox or chickenpox when spots are observed on his skin. However, it is precisely here where the high importance of combining an additional source of information is shown to be crucial.

We need to distinguish here what the patient *really* wants to know. He is interested in knowing what is the probability that he has the dreaded smallpox, after *spots have been observed* on his skin, i.e., x. In other words, he is worried that the value of $p(\theta_2|x)$ might be close to 1.

According to Byes' Rule:

$$p(\theta_2 \mid x) = \frac{p(x \mid \theta_2)\, p(\theta_2)}{p(x)} \tag{3.6}$$

Here $p(\theta_2)$ would be the "probability that smallpox occurs generally (in a human being)," which is termed the "prevalence" of a disease by the doctors. This, as it will be shown, is a *crucial* piece of *previous* knowledge. Numerically, it is determined by the probability that if we select at random a person from the general population of a region he or she will have smallpox. This is *previous* information that is known even before any data (presence of spots) is obtained from a specific situation (a specific patient). As such, this term would be called the *Prior Probability* for smallpox. Fortunately, nowadays that prior probability for smallpox is very low. For the example one could assume $p(\theta_2) = 0.001$. Similarly, the probability of a person (in the general population) displaying a pattern of spots on his/her skin (irrespective of cause) can be assumed to be $p(x) = 0.081$ for this example. This value would be substituted in the denominator of Bayes' Rule, and is customarily called the "*Marginal Likelihood*" (Stone 2015) or the "Evidence" (Candy 2009).

With the interpretation and the assumed values provided in the previous paragraph we can now use Bayes' Rule to find an answer to the *really* important question: What is the probability that the patient has smallpox, given that he displays skin spots as a symptom?:

$$p(\theta_2|x) = \frac{[0.9][0.001]}{[0.081]} = 0.011 \tag{3.7}$$

The good news is that, in this example, there is a probability of (about) one in a hundred that the spots seen in the patient are associated with smallpox. This probability, which is found "after" we have mutually enriched the knowledge from observed data, in this case $p(x|\ \theta_2)$, with prior knowledge, $p(\theta_2)$ is aptly called the "*Posterior Probability*."

In contrast, let's study how the combination of the *likelihood* for chickenpox $p(x|\ \theta_1) = 0.8$, with the previously known (general) prevalence of chickenpox (or *prior probability*) $p(\theta_1) = 0.1$, defines the *posterior probability* for having the less dangerous chickenpox, given that the patient displays skin spots:

$$p(\theta_1 \mid x) = \frac{p(x \mid \theta_1)\, p(\theta_1)}{p(x)} \tag{3.8}$$

which in this case is:

$$p(\theta_1|x) = \frac{[0.8][0.1]}{[0.081]} = 0.988 \tag{3.9}$$

By comparison, this result should reassure the patient: Given that he has spots on his skin, it is more likely that he has the less dangerous chickenpox ($p(\theta_1 \mid x) = 0.988$) than it is that he has smallpox ($p(\theta_2 \mid x) = 0.011$).

Notice that, the general format for Bayes' Rule used to obtain the POSTERIOR_PROBABILITY is:

$$[POSTERIOR\_PROBABILITY] = \frac{[LIKELIHOOD]\ [PRIOR\_PROBABILITY]}{[MARGINAL\_LIKELIHOOD]} \tag{3.10}$$

It should also be noticed that the main step in obtaining the much more informative POSTERIOR_PROBABILITY is chiefly achieved by *multiplying* the LIKELIHOOD with the PRIOR_PROBABILITY (Spagnolini 2018), and that the MARGINAL_LIKELIHOOD acts as just some type of "*normalizing element*" (Spagnolini 2018) (Candy 2009) that, actually, affected both posterior probabilities, $p(x|\theta_1)$ and $p(x|\theta_2)$, in the same way, and did not contribute as a distinguishing factor in their study. Because of this, Bayes' Rule is sometimes expressed in a broader, conceptual form as a proportionality (Spagnolini 2018) (Kovvali, Banavar, and Spanias 2013), instead of an equality, that relates the POSTERIOR_PROBABILITY to the *product* of the LIKELIHOOD times the PRIOR PROBABILITY, such as:

$$p(\theta_1|x) \propto p(x|\theta_1)\,p(\theta_1) \tag{3.11}$$

or

$$p(\theta_2|x) \propto p(x|\theta_2)\,p(\theta_2) \tag{3.12}$$

Or, in general terms:

$$[POSTERIOR\_PROBABILITY] \propto [LIKELIHOOD][PRIOR\_PROBABILITY] \tag{3.13}$$

and this is the interpretation that is used to apply Bayesian Estimation principles to the understanding of Kalman Filtering (Kovvali, Banavar, and Spanias 2013; Catlin 1988).

3.2 BAYES' RULE FOR DISTRIBUTIONS

Now we need to take one more step to generalize the use of Bayes' Rule to the context in which we will use it to understand the Kalman Filtering algorithm. In the previous example, we identified the occurrence of chickenpox as the random variable θ_1 and we identified the occurrence of smallpox as the random variable θ_2, and we studied them separately. However, we could also consider both of them as different values of a single random variable Θ, which then has possible values $\theta_1, \theta_1, \ldots \theta_D$.

$$\Theta = \{\theta_1, \ \theta_2, \ldots \ \theta_D\} \tag{3.14}$$

Here, the D values of Θ would be D different diseases (of whom θ_1 = chickenpox and θ_2 = smallpox are just the first 2). Then, the statistical characteristics of Θ would be studied considering the probability *distribution* for Θ.

Similarly, a variable X could represent S different symptoms:

$$X = \{x_1, \ x_2, \ldots \ x_S\} \tag{3.15}$$

In that case, for example, x_3 = skin spots would be only one value of the variable.

Just like we saw that the posterior probability of a patient having chickenpox when displaying skin spots, $p(\theta_1|x_3)$, and the probability of the patient having smallpox when displaying skin spots, $p(\theta_2|x_3)$, can be calculated separately, we can now see that the whole *Posterior Probability Distribution*, $p(\Theta|x)$, can be considered at once:

$$p(\Theta \mid x) = \frac{p(x|\Theta)\,p(\Theta)}{p(x)} \tag{3.16}$$

where:

- $p(\Theta|x)$ is the POSTERIOR_PROBABILITY_DISTRIBUTION, indicating the probability that each of the D diseases ($\theta_1, \theta_1, \ldots \theta_D$) is present, given that a symptom, x, (for example x_3 = skin spots) has been observed.
- $p(\Theta)$ is the PRIOR_PROBABILITY_DISTRIBUTION, indicating the probability that each of the D diseases ($\theta_1, \theta_1, \ldots \theta_D$) occurs in the general population (the "prevalences" of the diseases), without consideration of any specific symptom in any particular patient.
- $p(x)$ is the MARGINAL_LIKELIHOOD, indicating what is the chance (0 to 1) to find an individual at random that displays a particular symptom (for example x_3 = skin spots).
- $p(x|\Theta)$ is the LIKELIHOOD_FUNCTION, indicating what is the probability of observing symptom x (for example x_3 = skin spots) when a patient has each one of the D diseases in Θ: $\theta_1, \theta_1, \ldots \theta_D$. It should be noted that, while $p(\Theta|x)$ and $p(\Theta)$ *are* actually probability distributions (i.e., they *do* add up to 1.0), that is *not* the case for the LIKELIHOOD_FUNCTION (Stone 2015).

It should also be pointed out that the denominator of the right-hand side of Equation 3.16 will be a scalar if we are considering a single symptom (for example x_3 = skin spots). Once again, this will just act as a "normalizing element" (Haug 2012) (Candy 2009) and it plays a somewhat secondary role in the process of combining prior knowledge (the PRIOR_PROBABILITY_DISTRIBUTION) and specific evidence or data (the LIKELIHOOD_FUNCTION) to yield the important result, (i.e., the POSTERIOR_PROBABILITY DISTRIBUTION). Accordingly, Equation 3.16, i.e., Bayes' Rule for distributions, can be expressed more broadly as:

$$p(\Theta|x) \propto p(x|\Theta)\,p(\Theta) \tag{3.17}$$

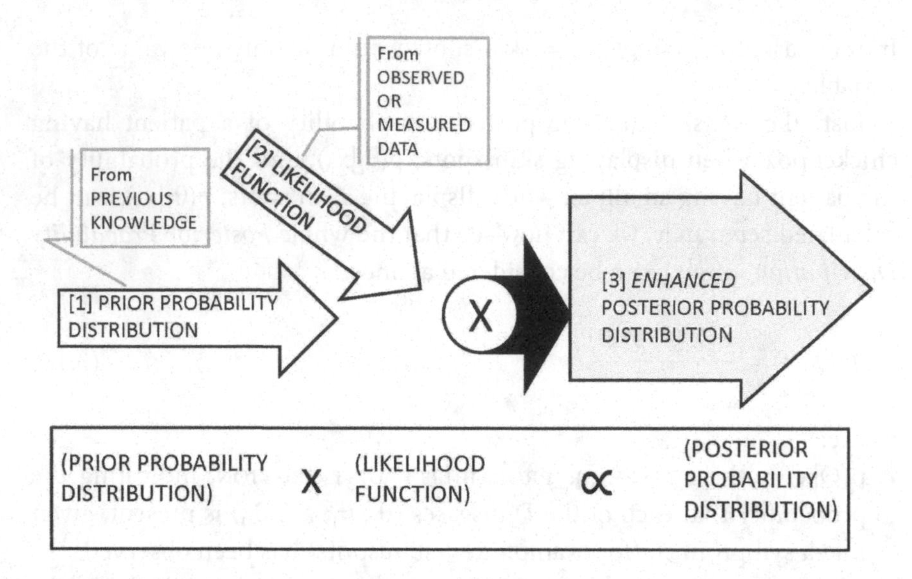

FIGURE 3.1 Graphical representation of the manner in which a Prior Probability Distribution is enriched with the information from a Likelihood Function (by multiplication), to yield an enhanced Posterior Probability Distribution. This is how the Bayesian Estimation process incorporates information from both sources.

Or, in general terms:

$$\begin{aligned}&[\text{POSTERIOR\_PROBABILITY\_DISTRIBUTION}]\propto\\&[\text{LIKELIHOOD\_FUNCTION}][\text{PRIOR\_PROBABILITY\_DISTRIBUTION}]\end{aligned} \tag{3.18}$$

Equation 3.18 is a powerful expression of a concept that will be key to our understanding of Kalman Filtering: A refined, "better informed" assessment of a probability distribution (the POSTERIOR_PROBABILITY_DISTRIBUTION) can be found by combining two partial assessments: A general, previous assessment (the PRIOR_PROBABILITY_DISTRIBUTION) with a second assessment more closely derived from the specific circumstances through data that is observed or measured (the LIKELIHOOD_FUNCTION). Further, the combination is performed, essentially, by multiplying one assessment by the other (Spagnolini 2018). To help summarize this important result, we present a graphical depiction of Equation 3.18 in Figure 3.1.

II

Where Does Kalman Filtering Apply and What Does It Intend to Do?

IN THIS PART OF the book we will discuss the Kalman Filter estimation approach and we will develop its algorithm equations through a series of reasoning steps that leverage the concepts reviewed in Part I. We start by presenting a minimal scenario where Kalman Filter can be applied. This first scenario will probably be easy to understand because all the key elements (the state variable, the measurement, etc.) are each represented by a scalar variable. In Chapter 5 we broaden these concepts and present the two general equations (model equation, measurement equation) that are required for the application of the Kalman Filter in a general, multivariate context, where the entities are represented by vectors and matrices. We also present a new scenario in which the system state is already represented by a 2-by-1 vector. Once the "components" involved in the Kalman Filter scenario have been identified, Chapter 6 presents a sequence of reasoning steps that the reader will be able to justify on the basis of the background developed in Part I, and which culminates with the statement of the prediction and correction equations for the Kalman Filtering algorithm. Chapter 7 completes this part of the book with reflections and discussions that highlight the benefits achieved by the Kalman Filtering algorithm and observations that we hope will help the reader develop an intuitive understanding of its components.

CHAPTER 4

A Simple Scenario Where Kalman Filtering May Be Applied

In this chapter we will present a very simple scenario that has the typical elements of an estimation problem of the kind Kalman Filtering is meant to help solve. We have tried to propose a system that will likely be familiar to any reader who has an engineering background, or even to a hobbyist or do-it-yourselfer. Our main goal is to help the reader in identifying, within this familiar situation, the concepts that play critical roles in the definition and implementation of the Kalman Filtering estimation process. These include state variable(s), measurement(s), system model(s), etc. We will emphasize that several of the variables in the scenario are, in fact, random variables, and, as such they have levels of uncertainty that must be expressed by appropriate statistical characteristics, such as their means and variances. We will also introduce the two governing relationships that are used to represent the model, the measurements and their interactions, employing the variable names that are commonly used when this kind of situation is addressed through Kalman Filtering.

4.1 A SIMPLE MODELING SCENARIO: DC MOTOR CONNECTED TO A CAR BATTERY

Imagine you buy a gizmo to be used in your car, which has for its objective to turn a rotary axis, slowly, at about 40 rotations per minute (This is

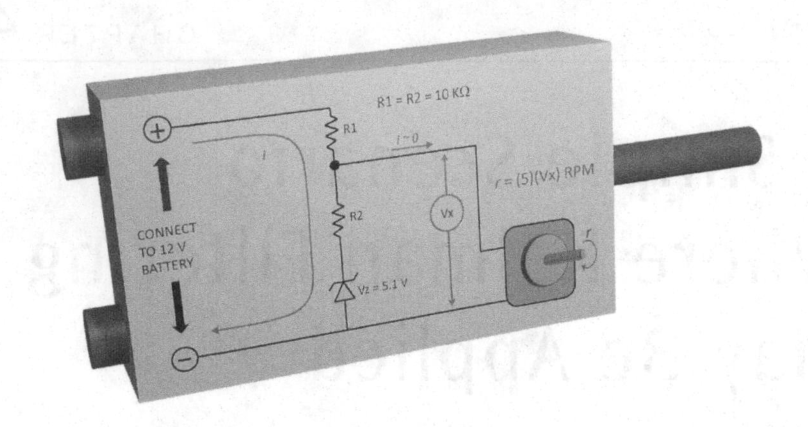

FIGURE 4.1 Diagram imprinted on the module.

just a hypothetical scenario and it does not refer to any actual gizmo in the market.) The gizmo comes in a plastic, sealed enclosure with only two electrical connection contacts labeled "+" and "-," and a sign that reads "Connect to 12 V battery." It also has the protruding metallic shaft that is supposed to turn. However, there is an image printed on the sealed module, as shown in Figure 4.1.

The diagram imprinted on our module indicates that it is meant to be powered by a car battery (nominal voltage = 12 VDC). It also indicates that the module comprises the series connection of R1, R2 and a "5.1 V" Zener diode (That is, the nominal reverse bias voltage is Vz = 5.1 V.) The module also indicates that the protruding shaft should be expected to turn at a speed "r" (in revolutions per minute), nominally defined to be 5 times the value of V_x, where V_x is the voltage at the node connecting R1 to R2, measured with respect to the negative contact. In our scenario, we will assume that we would like to estimate, as well as we can, what the value of that V_x is.

In this scenario V_x plays the role that we will call "state variable." That is, it is a variable that characterizes the state of operation of the system (our gizmo), but is, however, not directly accessible for us to measure. In general, we use the letter "x" with the use of subindices (if necessary) to identify these "state variables." Further, in general, we should allow for the possibility that these state variables can vary through time, so we can identify for our scenario:

$$V_x = x_1(t) \tag{4.1}$$

We should clarify that this book addresses the case where all the variables are considered *discrete-time (sampled) variables.* That is, in spite of using the variable "t" we will be referring throughout to *sequences of values that are "digitized," being recorded only once every "sampling interval,"* (for example, every 1 millisecond). This will be the case for all variables that we will discuss. Any variable, written in this book as, for example, "j(t)," represents a discrete-time sequence. We use "t" as discrete-time index to keep compatibility with other texts that discuss Kalman Filtering of discrete-time signals.

It is very important to realize that our scenario clearly offers *two ways* to try to estimate V_x at any time:

1. The scenario implies a *model* which could *predict* the value of V_x, at any time, based on our knowledge of elementary electrical circuits.
2. We could also *measure* the effective rotational speed of the protruding shaft, *r*, and, based on the relationship printed in the module ("r = 5 V_x"), we could define the corresponding value of V_x. So, we could try to define V_x from knowledge of the rotational speed measurements, *r*(t).

Let's look closer into each one of those possibilities.

4.2 POSSIBILITY TO ESTIMATE THE STATE VARIABLE BY PREDICTION FROM THE MODEL

A model for $x_1(t)$ could be derived considering the voltage "division" established by R1 and R2:

$$i = \frac{VBATT - Vz}{R1 + R2} \tag{4.2}$$

$$V_x = Vz + iR2 = Vz + \frac{(VBATT - Vz)R2}{R1 + R2} \tag{4.3}$$

If we were to substitute the *nominal* values of the elements:

$$V_x = (5.1V) + \frac{(12V - 5.1V)(10{,}000)}{10{,}000 + 10{,}000} \tag{4.4}$$

$$V_x = 8.55V \tag{4.5}$$

$$x_1(t) = V_x = 8.55V \tag{4.6}$$

Here, as the model shown in the diagram does not indicate changes through time:

$$x_1(t+1) = x_1(t) \tag{4.7}$$

Therefore, V_x is expected to remain at 8.55 V. While this is the "basic" description of the relationships between variables in the system, we need to recognize now the multiple sources of *uncertainty* all around this scenario.

4.2.1 Internal Model Uncertainty

A critical component of the "basic" model is the use of "nominal" values for VBATT, Vz, R1 and R2. However, we know that resistors have a certain "tolerance," which *warns us* that their *effective value may actually be* as much as 10% above or below the nominal value (if the resistors have a silver 4th band). The actual voltage between the terminals of the Zener diode depends on the reverse current through it (which, in turn could be impacted by departure from the nominal values of R1, R2 and VBATT). Further, R1, R2 and Vz are known to change with temperature variations. Additionally, in a real-life case (e.g., if VBATT is the voltage of the battery in a car), the switching on and off and modifications in the current demanded by other devices connected to the same battery could introduce up-and-down variations in the effective value of VBATT ("noise"). So, although we have the model described by Equation 4.7, we know that the value (8.55 V) predicted by this model for the "next" time (which in this case would just keep it at 8.55 V), may be different from the "true value" by a certain amount. Because of the presence of uncertainties in the model itself, there can be differences (positive or negative, possibly alternating) between the "true value" of the state variable and the value that would be predicted by the model.

4.2.2 External Uncertainty Impacting the System

In addition, external phenomena, such as electromagnetic interference, may cause noise voltages to appear superimposed in the node between R1 and R2. In a general scenario, these external influences could even include "external inputs" (or "control inputs") that are voluntarily applied to influence how a system changes from one state to the other. Even if no value is applied voluntarily, it is very likely that external events (such as the electromagnetic noise mentioned previously) may have an impact on the evolution of the system from one state to the next. Therefore, we should always include these external contributions to the uncertainty that will exist in the prediction of the value(s) of the state variable(s) from one sampling instant to the next.

4.3 POSSIBILITY TO ESTIMATE THE STATE VARIABLE BY MEASUREMENT OF EXPERIMENTAL VARIABLES

If, in contrast, we were to rely on the measurement of the rotational speed of the shaft as the means to establish the value of the state variable at any time:

$$r(t) = 5x_1(t) \tag{4.8}$$

$$x_1(t) = \frac{r(t)}{5} \tag{4.9}$$

In this case r(t) will be a sequence of rotational speed measurements that we can obtain from some form of measurement system every sampling interval, and they will enable us to obtain measurement-based estimates of $x_1(t)$ with the same periodicity.

4.3.1 Uncertainty in the Values Read of the Measured Variable

With respect to the measurement of rotation speed values, we need to acknowledge that it is very likely that our recorded values will also be affected by "error." For example, if we use a chronometer to measure 1 minute and attempt to count the revolutions completed by the axis of the motor in that interval we could fail to stop the count exactly after 1 minute. We could also over- or underestimate the final portion of the last revolution observed in 1 minute. Even if an instrument is being used to obtain the values of the measured variable, we know that such measurements are also frequently distorted by random noise and instrumental inaccuracies. So, we need to be prepared to accept that we will have a measurement of r that may be above or below the true value of rotation speed. That is, we need to include in our study a representation of the uncertainties that exist in recording the variables that we will measure.

Under the assumptions explained in the earlier paragraphs, this univariate scenario is appropriate for the application of Kalman Filtering to obtain a better estimate of the single state variable, $V_x = x_1(t)$, than we could obtain from the model alone or than we could derive from the measurements of rotation speed of the shaft alone.

The next chapter presents the general framework proposed for the Kalman Filter to estimate variables in systems which have a *model* with internal and external uncertainties and a mechanism for obtaining *measurements* (also with uncertainties) from the system.

After the reasoning leading to the Kalman Filter algorithm is presented (Chapter 6), Chapter 9 will show a specific application of Kalman Filtering to *this scenario*, using the MATLAB® implementation of the algorithm introduced in Chapter 8.

CHAPTER 5

General Scenario Addressed by Kalman Filtering and Specific Cases

Now that we have shown the reader a very simple, practical scenario with the elements that are typical of a Kalman Filtering situation (a model with uncertainties, and measurements with uncertainties), we will present the reader with the general estimation challenge that Kalman addressed in his seminal 1960 paper. This challenge will be presented (as Kalman did) through two key equations (The state transition, or model, equation and the measurement equation) and, also, as a block diagram. This will help to identify the elements, parameters and interrelations that exist in the estimation challenge we will address, as well as the assumptions made about them. We will then express the simple practical scenario described in the previous chapter as an instance of the general estimation challenge. Afterwards we will also cast the estimation of the height of a falling object through time in the generic estimation model. This second situation is already a closer case to the actual use of Kalman Filtering estimation in engineering practice.

5.1 ANALYTICAL REPRESENTATION OF A GENERIC KALMAN FILTERING SITUATION

Here we will describe a generic context that represents situations that contain the key elements that we emphasized in the previous battery-Zener-motor example: *There is a system with state variables and a model that allows us to predict the "next values" of the state variables and, in addition, there are measurements that can be taken from the same system, which also provide information that can lead to estimates of the state variables.*

In his 1960 paper (Kalman 1960), Kalman proposed his approach to the "Solution of the Wiener Problem," which seeks to provide the best possible (recursive) estimate of the state variables collected on a vector **x**(t) in the context represented by these two equations (where the variables in bold are matrices or vectors):

$$\boldsymbol{x}(t+1)=\boldsymbol{\Phi}(t+1;t)\boldsymbol{x}(t)+\Delta(t)\boldsymbol{u}(t) \tag{5.1}$$

$$\boldsymbol{y}(t)=\mathbf{M}(t)\boldsymbol{x}(t) \tag{5.2}$$

These two equations represent the "general linear discrete-dynamic system" that also appears in Kalman's paper (Kalman 1960) as a block diagram in Figure 2 of his paper, shown here as Figure 5.1.

Equation 5.1 (state transition or "model" equation) is represented by the left and middle regions of Figure 5.1, and it summarizes our ability to predict the "next" value of the state variable(s) through a model of the situation. (In our previous simple example, our model would predict that the "next" state variable value would simply be the same as the previous value: **x**(t + 1) = **x**(t).) The loop at the center of Figure 5.1 represents the basic

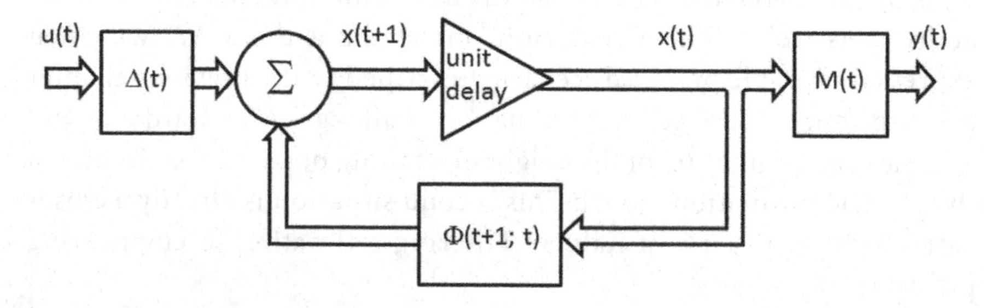

FIGURE 5.1 "General linear discrete-dynamic system" for which the state, x(t), will be estimated. (Figure 2 from Kalman 1960. Used with permission from ASME.)

evolution of the state variables of the system, from **x**(t) to **x**(t + 1). The "self-evolution" of the system is represented by the "state transition matrix," **Φ**(t + 1;t), which is used to pre-multiply the vector **x**(t). However, the new *predicted* values of the state vector, **x**(t + 1) *must also consider* the impact of external influences, represented by the product of a vector **u**(t) pre-multiplied by matrix **Δ**(t), which transforms the values in **u**(t), sometimes called "control inputs," into corresponding state variable contributions, such that they can be added to the product of **Φ**(t + 1;t) by **x**(t), to finally yield **x**(t + 1). In some contexts (e.g., automatic control literature), **u**(t) is often used to represent commands applied to drive the system to desired states. However, in our context, **u**(t) may represent an input applied on purpose to "drive" the system, but it may also represent only *external sources of uncertainty (noise)* impacting the state variables.

In fact, in his 1960 paper (Kalman 1960), Kalman makes the following assumptions about the nature and characterization of ***u***(t):

"{ ***u***(t) } is a vector-valued, independent, gaussian random process, with zero mean, which is completely described by (in view of Theorem 5(C))

E **u**(t) = **0** for all t;

E **u**(t)**u'**(s) = **0** if t ≠ s

E **u**(t)**u'**(t) = **Q**(t)"

where ***E*** is the expectation operator and **u'**(t) is the transpose of **u**(t) (Kalman specifically states that, in the paper, the transpose of a matrix "will be denoted by the prime.") Therefore, **Q**(t) is the covariance matrix for **u**(t).

Similarly, Kalman's paper assumes that **x**(t) is also a vector that has a Gaussian pdf in developing the "Solution of the Wiener Problem," which he poses in the following terms:

"Let us now define the principal problem of the paper.

Problem I. Consider the dynamic model:

$$\boldsymbol{x}(t+1)=\boldsymbol{\Phi}(t+1;t)\boldsymbol{x}(t)+\boldsymbol{u}(t)$$
$$y(t)=\mathbf{M}(t)\boldsymbol{x}(t)$$

where **u**(t) is *an independent gaussian random process of n-vectors with zero mean,* **x**(t) is *an n-vector,* **y**(t) is *a p-vector* ($p \leq n$), **Φ**(t + 1;t), **M**(t) *are* $n \times n$, *resp.* $p \times n$, *matrices whose elements are nonrandom functions of time. Given the observed values of* **y**(t_0), . . . , **y**(t) *find an estimate* **x***(t_1|t) of **x**(t_1) *which minimizes the expected loss.* (See Figure 2, where **Δ**(t) = **I**.)"

Note: Figure 2 in Kalman's paper is Figure 5.1 in this chapter.

This is also the estimation problem that we consider in this book.

In our study, the second equation that defines the scenario (Equation 5.2), known as the "measurement equation," is also an essential part of the context in which the estimation solution will be sought. Equation 5.2 is illustrated by the right region of Figure 5.1, and represents the fact that we frequently may not be able to record the state variables, **x**(t), directly and will have to "observe" other variables or MEASUREMENTS, **y**(t) to monitor the performance of the system. (In our simple example, we know that the rotational speed (in RPMs) that we can observe in the shaft will be, numerically, 5 times the value of the V_x node voltage, in Volts.) In Equation 5.2, the matrix **M**(t) establishes the conversion from state variables to measurements. (Like the factor "5" converts V_x volts to RPMs in the rotating shaft.)

In this book we will study the same situation, but *we will use a slightly different nomenclature that matches many newer texts and articles* (The equivalences should be obvious: **Φ** is substituted by **F**; **Δ** is substituted by **G**; **M** is substituted by **H**; **y** is substituted by **z**):

$$\boldsymbol{x}(t+1) = \mathbf{F}(t)\boldsymbol{x}(t) + \mathbf{G}(t)\boldsymbol{u}(t) \tag{5.3}$$

$$\boldsymbol{z}(t) = \mathbf{H}(t)\boldsymbol{x}(t) \tag{5.4}$$

Also, notice that we have simplified the notation for the "state transition matrix," from **Φ**(t + 1;t) to simply **F**(t). The more explicit notation for **Φ**(t + 1;t) clearly indicates that the transition towards the state predicted for t + 1 will be made on the basis of the current characteristics of the system (at time t). We should understand the same from the more compact notation **F**(t).

To begin gaining a feeling for how these matrices and vectors are actually formed, let's consider, just as a hypothetical example, a system with two state variables, x_1 and x_2, two external input variables, u_1 and u_2, and three measured variables z_1, z_2 and z_3. Under those conditions, Equations 5.3 and 5.4 would look like this:

$$\begin{bmatrix} x_1(t+1) \\ x_2(t+1) \end{bmatrix} = \begin{bmatrix} f_{11} & f_{12} \\ f_{21} & f_{22} \end{bmatrix} \begin{bmatrix} x_1(t) \\ x_2(t) \end{bmatrix} + \begin{bmatrix} g_{11} & g_{12} \\ g_{21} & g_{22} \end{bmatrix} \begin{bmatrix} u_1(t) \\ u_2(t) \end{bmatrix} \quad (\textit{example of Eq. } 5.3)$$

$$\begin{bmatrix} z_1(t) \\ z_2(t) \\ z_3(t) \end{bmatrix} = \begin{bmatrix} h_{11} & h_{12} \\ h_{21} & h_{22} \\ h_{31} & h_{32} \end{bmatrix} \begin{bmatrix} x_1(t) \\ x_2(t) \end{bmatrix} \quad (\textit{example of Eq. } 5.4)$$

Equation 5.3 represents the SYSTEM MODEL, encoded in matrix **F**, which will predict the next state of the system, **x**(t + 1), given the current state, **x**(t), and also accounts for how the prediction is impacted by external influences, (e.g., external noise and/or purposefully applied inputs), through the product **G**(t) **u**(t). So, if we knew the values of the state variables at a given time (let's say t = 0), and had (for example) a constant **F**(t) and assume (for the time being) that we could neglect the external inputs and noise), we should be able to recursively compute the values of the state variables for t = 1 (that is, **x**(1)) and, using those results, proceed to obtain **x**(2), and, with those results, calculate **x**(3), etc. In other words, this system model equation *already* provides us with a first estimate for the state variables at any time. However, please note that if we were to "blindly believe" the predicted state estimate obtained from the model alone, we might be getting values that could be *very different* from the *real* values of the state variables that the physical system is going through, and we would not even know it!

So, one way to begin doing something about this is to acknowledge that the state vector obtained from the model alone, at every iteration, brings along with it some level of uncertainty. A good way to represent this with respect to all the state variables (and their inter-relationships) is through the *state covariance matrix* **P**(t). A covariance matrix, **P**(t), for the random vector **x**(t), is succinctly defined (Childers 1997), as:

$$\boldsymbol{P}(t) = E\left\{\left[\boldsymbol{x}(t) - \boldsymbol{\mu}_x(t)\right]\left[\boldsymbol{x}(t) - \boldsymbol{\mu}_x(t)\right]^T\right\} \tag{5.5}$$

where the expectation operator E{ } can be interpreted as obtaining the mean of the quantity to which it is applied. In Equation 5.5, $\boldsymbol{\mu}_{\mathbf{x}}(t)$ is a vector of the same size as **x**(t), containing the means of the state variables, that is: $\boldsymbol{\mu}_{\mathbf{x}}(t)$ = E{**x**(t)}. Therefore, we see that **P**(t) is the covariance matrix for the vector **x**(t), just as defined back in Chapter 2, where Equations 2.5, 2.6 and 2.7 exemplified the covariance matrix and Equation 2.8 provided its definition.

When we ask specifically what will be the uncertainty of the PREDICTED state vector, **x**(t + 1), we must recognize that the contribution of **x**(t) to **x**(t + 1) is obtained by pre-multiplying it by **F**(t). In other words, **x**(t + 1) results, in part, from applying a linear transformation to **x**(t). Therefore, the covariance of the state estimate, **x**(t + 1) is, in part, defined by the modification of **P**(t) we learned when considering multivariate random variables undergoing a linear transformation: i.e., it will be transformed into **F**(t) **P**(t) $\mathbf{F}(t)^T$ (recall Equation 2.12). However, the full uncertainty of the overall estimated **x**(t + 1) will also include the uncertainty introduced by external factors (noise), represented by the covariance matrix **Q**(t) (Assuming for simplicity G is the identity matrix, as Kalman suggested by proposing "Δ(t) = I" for his Figure 2, which is practical if no purposeful command inputs are being considered):

$$\mathbf{Q}(t) = E\left\{\left[\boldsymbol{u}(t) - \boldsymbol{\mu}_u(t)\right]\left[\boldsymbol{u}(t) - \boldsymbol{\mu}_u(t)\right]^T\right\} \tag{5.6}$$

Then the uncertainty of the predicted state vector at time t + 1, defined on the basis of the uncertainty of **x**(t), will be:

$$\boldsymbol{P}(t+1) = \boldsymbol{F}(t)\boldsymbol{P}(t)\boldsymbol{F}(t)^T + \boldsymbol{Q}(t) \tag{5.7}$$

Notice that this equation provides a recursive mechanism to "track" the uncertainty of the state estimates as time goes on.

However, just knowing that **x**(t) has an uncertainty associated with it, even if we can track the progress of that uncertainty, would not really improve the estimates obtained from the model in a recursive manner. Awareness of the fact that the uncertainty of the state estimates we obtain from the model may keep growing is not enough. We would like to be able to intervene, correcting the estimate, but that requires an additional source of information that can guide the way in which the estimates should be corrected.

The second equation that completes the description of the scenario (Equation 5.4) indicates that we have *a second source of information* to attempt an estimation of the estate variables. We are, in fact, constantly obtaining practical measurements collected in the measurement vector, **z**(t), which are associated with the state vector, **x**(t), according to that second equation:

$$\boldsymbol{z}(t) = \mathbf{H}(t)\boldsymbol{x}(t) \tag{5.8}$$

This equation suggests that we could find a "practical" (as opposed to model-based) evaluation for the values in **x**(t) by plugging in the measurements **z**(t). The measurements in the vector **z**(t) will, indeed, be taken into account for our final definition of the "best estimate" of the state variable vector. But this second approach is (unfortunately) *also imperfect.* As we have mentioned before, most practical measurements will typically be plagued with "measurement uncertainty" or "measurement error." We will model this uncertainty through **R**, the covariance matrix for vector **z**(t):

$$\boldsymbol{R}(t) = E\left\{\left[\boldsymbol{z}(t) - \boldsymbol{\mu}_{\boldsymbol{z}}(t)\right]\left[\boldsymbol{z}(t) - \boldsymbol{\mu}_{\boldsymbol{z}}(t)\right]^{T}\right\} \tag{5.9}$$

The Kalman Filter process assumes that **z**(t) is also a vector with Gaussian pdf.

5.2 UNIVARIATE ELECTRICAL CIRCUIT EXAMPLE IN THE GENERIC FRAMEWORK

So, how does our simple example involving a 12-volt battery, a couple of resistors, a Zener diode and a motor (Figure 4.1) fit in the generic framework described in Equations 5.3 and 5.4?

The first thing to realize is that our state variable vector x(t) is actually a scalar. We only have one state variable, $x_1(t) = V_x(t)$, therefore:

$$\boldsymbol{x}(t) = x_1(t) \tag{5.10}$$

Because of this, Equation 5.3 will be a scalar equation in this extremely simple case. Because the circuit is fixed and even the nominal value of VBATT is constant, the model would predict that the state variable will hold a constant value:

$$x_1(t+1) = x_1(t) \tag{5.11}$$

Therefore, this means **F**(t) = [1] = 1.

On the other hand, we recognize that there is uncertainty in the ability of our model to predict the next state from the current state, and we are encoding this acknowledgement in **P**(t), which is the covariance matrix of vector **x**(t). Having only one state variable, $x_1(t)$, P(t) will be a scalar given by the variance of $x_1(t)$:

$$\mathbf{P}(t) = \sigma^2_{x1} \tag{5.12}$$

We have said already that, in many cases, we will not apply control inputs, but we will recognize that there can be external noise affecting the evolution of the states of the system. If the external noise is represented by a Gaussian random distribution with 0 mean and variance σ^2_u, then the complete substitution in Equation 5.3 yields:

$$x(t+1) = x(t) \tag{5.13}$$

with a predicted variance for $x_1(t + 1)$ calculated as:

$$P(t+1) = [1]\sigma^2_{x1}[1]^T + \sigma^2_u \tag{5.14}$$

$$P(t+1) = \sigma^2_{x1} + \sigma^2_u \tag{5.15}$$

Okay, but, how do we determine the values that should be substituted for σ^2_{x1} and σ^2_u in this example? Unless we have an analytical model for the uncertainties involved in the upgrade of the state variables and parameters to characterize the external noise, we may need to just propose values for σ^2_{x1} and σ^2_u, keeping some of these points in mind:

1. In this context, σ^2_{x1} and σ^2_u represent the level of "mistrust" that we have in the correctness of the model and in the isolation of our system from external noises. If we are not confident in either one of these, we may assign "larger" values to these variances. (In particular, consider that σ^2_u is a direct representation of the "power" of the noise being added into the system.)
2. Observe that **Q**(t) will be the same second term in Equation 5.7, remaining unchanged through the iterations. We will see later that, as the Kalman Filter evolves, the covariance matrix of the state vector, **P**(t), may change, however the covariance matrix of the external noise, **Q**(t), is not being affected by the Kalman Filter processing. (It is, after all, the characterization of *external* elements.) Therefore, **Q**(t) will often act as a "lower bound" of the "weight or importance" of **P**(t) during the Kalman Filter processing. This "weight or importance" of **P**(t) will be a significant characteristic of the Kalman Filtering process, as we will describe in later sections. For this example **Q**(t) is simply σ^2_u (constant).

And, how about the measurement equation? In this case there is only one variable being measured: r (rotation speed in revolutions per minute).

Therefore, **z**(t) = r(t), and Equation 5.4 will also reduce to a scalar equation, where **H**(t) will simply be the conversion factor from x_1 in volts to r in RPMs (which is constant). So, **H**(t) = 5, and, therefore:

$$z(t)=r(t)=(5)(x_1(t)) \tag{5.16}$$

The "experimental" uncertainty for the measurement is represented by the covariance matrix, **R**(t), for the measurement vector **z**(t). In our example, where **z**(t) = r(t), this simplifies to **R**(t) = σ^2_r.

The value(s) included in the covariance matrix **R**(t) may be:

1. Approximated on the bases of the specifications and/or calibrations of the instruments used (which are sometimes provided by their manufacturers);

   or

2. Since we have started from the assumption that we do have access to the measurement variables, one could, conceivably, record some of that data in advance, and use those preliminary data to evaluate the variances and covariances between the measurement variables. In the practical example we are considering, perhaps one could power the circuit and take multiple measurements of the RPMs so that σ^2_r could be approximated.

In summary, at any sampling instant, we will have two sources of information to try to define our best estimate of the current set of state variables: Equation 5.3 will provide an explicit suggested estimate for **x**(t + 1), coming from the model, along with its level of uncertainty, encoded in **P**(t + 1). On the other hand, we will also have the most recent set of samples from the measurement variables, **z**(t), which (implicitly) suggest, through Equation 5.4, what the values in **x**(t) should be. And we will also have the level of uncertainty of those **z**(t) values, encoded in **R**(t). Chances are, these two sources of information will not point to the same estimate for the state variables. Who are we to believe? The Kalman Filtering approach provides an answer to this daunting question that is, at once, (remarkably) well-aligned with common sense, and theoretically optimal, from different perspectives (Grewal and Andrews 2008; Maybeck 1979; Anderson and Moore 2012; Gelb et al. 1974; Kovvali, Banavar, and Spanias 2013; Serra 2018; Simon 2001). The rationale and step-by-step instructions for the implementation of this answer will be discussed in Chapter 6.

5.3 AN INTUITIVE, MULTIVARIATE SCENARIO WITH ACTUAL DYNAMICS: THE FALLING WAD OF PAPER

From very early in history, trying to predict the way in which a body would move under "free fall" conditions was an appealing challenge for inquisitive people. According to legend (Drake 1978) Galileo performed studies and experiments dropping objects from the top of the Leaning Tower of Pisa. Such scenario, however, would not be ideal for the study of "free fall," since any object dropped from the height of the tower will interact with the intervening air mass present in the trajectory. In particular, the friction experienced by the falling body, which will be related to its cross section, will result in a force which opposes the movement of the object and, as such, slows the fall.

If the experiment is performed, instead, in a giant vacuum chamber, where the air has been extracted, as Brian Cox and Andrew Cohen describe in their book *The Human Universe* (Cox and Cohen 2014), the effect of friction would be removed and the free fall process will only be driven by the gravitational acceleration. (In that case both a bunch of feathers and a bowling ball dropped simultaneously will reach the ground at the same time, as shown in Episode 4 of the BBC video series "Human Universe," presented by Dr. Cox.)

If, at least initially, we do not consider the air friction in the study of the evolution of the height above ground level (y) through time when an object is dropped from an initial height y_0, the only acceleration at play is gravity ($g = 9.81\ m/s^2$). Therefore, the second derivative with respect to time of the height, (denoted with double prime), would be:

$$y''(t) = -g \tag{5.17}$$

We will first just study the movement of an object along an axis "y" but subjected to a constant POSITIVE acceleration, a_y, as it moves in the POSITIVE direction of the y axis. Then, we will just substitute $a_y = -g$ to obtain the equations that truly apply to the free falling case.

We will proceed to the study of the changes of y position and its derivatives in small time intervals, Δt, (as would be the sampling interval in a discrete-time system). For each interval of study we will assume there is a constant acceleration, a_y. In this case, the rate of change of the speed (denoted with a single prime), from the beginning ($t - \Delta T$) to the end (t) of a ΔT interval, can be approximated as:

$$y''(t) = \frac{y'(t) - y'(t - \Delta t)}{\Delta t} = a_y \tag{5.18}$$

From this:

$$y'(t) - y'(t - \Delta t) = a_y \Delta t \tag{5.19}$$

and

$$y'(t) = y'(t - \Delta t) + a_y \Delta t \tag{5.20}$$

Now, we can imagine the change in y position (from time t – ΔT to time t), as consisting of two components: A first component, "C1" caused by the previous velocity, y′(t – ΔT), over the interval of observation, ΔT, and a second component "C2," which is the "additional" change of position due to the fact that the velocity itself changes during the observation interval. To visualize this we could graph the way in which the velocity changes from time t – ΔT to time t, as in Figure 5.2:

Since Figure 5.2 displays the evolution of the speed, y′, over the interval of observation, ΔT, then "the area under the curve" is the change of position. We can see here that the change of y position will have the two components, C1 and C2, which we mentioned before. We also notice that the corresponding areas can be described as a rectangle and a triangle. As

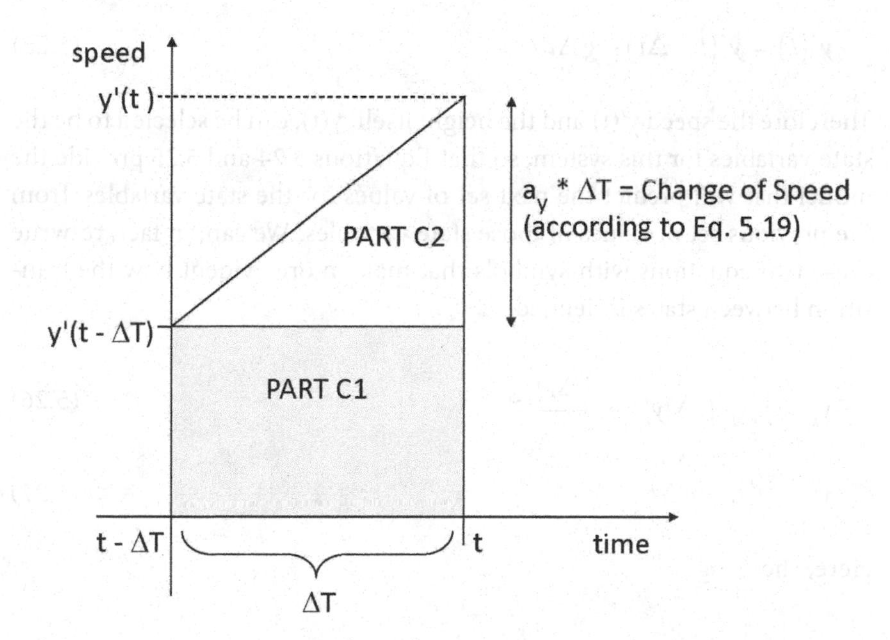

FIGURE 5.2 Derivation of the kinematic equation at constant acceleration.

such, we can use the formulas for the area of a rectangle and the area of a triangle to calculate the change of y position:

$$\begin{aligned} y(t) - y(t-\Delta t) &= [AreaC1] + [AreaC2] \\ &= [\Delta t y'(t-\Delta t)] + \left[\frac{(\Delta t a_y)(\Delta t)}{2}\right] \end{aligned} \tag{5.21}$$

$$y(t) - y(t-\Delta t) = [\Delta t y'(t-\Delta t)] + \left[\frac{(\Delta t)^2 a_y}{2}\right] \tag{5.22}$$

$$y(t) = y(t-\Delta t) + \Delta t y'(t-\Delta t) + \frac{(\Delta t)^2 a_y}{2} \tag{5.23}$$

For the study of a free-falling object, all we need to do is consider the y axis pointing UP with its origin at ground level, and substitute $a_y = -g$ in Equations 5.20 and 5.23. Then, the equations for the instantaneous height, y(t), and for its rate of change, y′(t) will be:

$$y(t) = y(t-\Delta t) + \Delta t y'(t-\Delta t) - \frac{(\Delta t)^2 g}{2} \tag{5.24}$$

$$y'(t) = y'(t-\Delta t) - g\Delta t \tag{5.25}$$

Therefore the speed y′(t) and the height itself, y(t), can be selected to be the state variables for this system, so that Equations 5.24 and 5.25 provide the model that will predict the next set of values for the state variables, from the previous set of values of those state variables. We can, in fact, re-write these two equations with symbols that make more evident how the transition between states is defined:

$$y_k = y_{k-1} + \Delta t y'_{k-1} - \frac{(\Delta t)^2 g}{2} \tag{5.26}$$

$$y'_k = y'_{k-1} - g\Delta t \tag{5.27}$$

Here, choosing

$$\boldsymbol{x} = \begin{bmatrix} y_k \\ y'_k \end{bmatrix} \tag{5.28}$$

$$\boldsymbol{F} = \begin{bmatrix} 1 & \Delta T \\ 0 & 1 \end{bmatrix} \tag{5.29}$$

$$\boldsymbol{G} = \begin{bmatrix} -\left(\frac{1}{2}\right)(\Delta T)^2 \\ -\Delta T \end{bmatrix} \tag{5.30}$$

and substituting $\mathbf{u}(t) = g$, we can express the two equations that define the model in the format we proposed for state estimation through Kalman Filtering:

$$\boldsymbol{x}(t+1) = \boldsymbol{F}(t)\boldsymbol{x}(t) + \boldsymbol{G}(t)\boldsymbol{u}(t) \tag{5.31}$$

$$\begin{bmatrix} y_{k+1} \\ y'_{k+1} \end{bmatrix} = \begin{bmatrix} 1 & \Delta T \\ 0 & 1 \end{bmatrix} \begin{bmatrix} y_k \\ y'_k \end{bmatrix} + \begin{bmatrix} -\left(\frac{1}{2}\right)(\Delta T)^2 \\ -\Delta T \end{bmatrix} g \tag{5.32}$$

But, in reality, as mentioned before, the fall will not be driven exclusively by the acceleration of gravity, $g = 9.81\ m/s^2$. If we imagine, specifically, that the object we release from an initial height, y_0, is a big paper wad, like the one we create when we dispose of the wrapping paper we get with a gift, then its fall will also be affected by the resistance that the air opposes to its falling motion. Further, taking into account the typical irregular crumpling pattern of a wad of paper, and the fact that it is likely to rotate as it falls, then the amount of air friction opposing the fall will likely not be constant. Instead, we may suggest modeling that opposing force as having a mean value associated with a mean opposing acceleration (which we will call g_{back} to denote its opposition to gravity). Then, we can model the instantaneous opposing acceleration as varying randomly, above and below g_{back}, according to a Gaussian distribution. Therefore, to account for this, in our model, we will substitute g (constant) for a time series, that we have called "actualg," characterized by a Gaussian distribution with mean $g - g_{back}$, which will have a standard deviation of value $\sigma_{actualg} = gsd$ (gsd: standard deviation of the "actualg(t)" time series).

In this case, the initial values of the state variables will be the initial speed (which we would assume is 0 m/s), and the initial height from which we drop the paper wad, y_0. Those initial estimates of the two state variables, however, will have some level of uncertainty, which would need to be specified in the initial covariance matrix, $\mathbf{P}_0$.

These considerations identify all the elements necessary in the model for the phenomenon, which provides estimates of both height and speed as the paper wad falls.

We can, in addition, assume that we have an instrument (e.g., a laser rangefinder) that can actually provide empirical measurements of the height, which we will identify as the sequence z(t), at the same time increments, ΔT, which we are using for the implementation of the model. Because the height is the first variable in the state variable vector, the relationship between the (scalar) measurement, z(t), and the state variable vector for this case is:

$$z(t) = \boldsymbol{H}\boldsymbol{x}(t) = [1,0]\begin{bmatrix} y \\ y' \end{bmatrix} \tag{5.33}$$

In this case, the uncertainty for the measurement, generally represented by the covariance matrix **R**, will be just a scalar, given by the variance of the height measurements provided by the instrument. This is a parameter that would normally be determined on the basis of the principle of operation and characteristics of the instrument, sometimes provided by the manufacturer.

In this chapter, we have presented the general setup for state variable estimation through Kalman Filtering, represented by Equations 5.3 and 5.4. Further, we re-visited the univariate scenario presented in the previous chapter and identified the variables and parameters of that situation within the general framework of Equations 5.3 and 5.4. Then we introduced a second scenario where the goal is to obtain the estimate of the height (and the speed) of a falling paper wad, taking into account the resistance presented by the air to the movement of the paper. This was already a case where several of the variables and parameters in the general framework of Equations 5.3 and 5.4 were matrices or vectors, that is, a multivariate case. We will present and study Kalman Filter solutions to these estimation problems in Chapters 9 and 10, where we will use the MATLAB® implementation of the Kalman Filter algorithm that we will develop in Chapter 8.

CHAPTER 6

Arriving at the Kalman Filter Algorithm

In this chapter we will use background concepts presented in Chapters 1–3 to follow a reasoning sequence that will lead us to the step-by-step algorithm for the implementation of the Discrete Kalman Filter. This reasoning sequence may not qualify as a "strict proof" of the algorithm, and that is not its purpose. Instead, the objective of following the reasoning from previous background knowledge (which we hope the reader already feels comfortable with) to the algorithm listing is to foster in the reader a convinced assimilation of the considerations that justify the appropriateness of the algorithm. We believe that this process will, simultaneously, develop an *intuitive understanding* of the terms and parameters that are part of the algorithm. This, in turn, will help the reader identify them in any practical prospective application of the Kalman Filtering process.

6.1 GOALS AND ENVIRONMENT FOR EACH ITERATION OF THE KALMAN FILTERING ALGORITHM

We will start the reasoning that will lead us to the equations that define the Kalman Filtering algorithm by re-stating the circumstances in which it will operate and the goal that it pursues.

We are studying a dynamic system, whose behavior is characterized by the value of some variables grouped in the state vector **x**(t). We have a model that (perhaps imperfectly) describes the evolution of those state variables, recurrently, from one discrete-time index, t, to the next one, t + 1. Further, the model is expressed as a state transition matrix, **F**(t), so that the "self evolution" of the state at time t to the state at time t + 1 is obtained by simple pre-multiplication of **x**(t) by **F**(t). However, we also have provisions to represent how other variables, which are perhaps manipulated on purpose, contained in the "input vector," **u**(t), may also influence the next state. The influence of **u**(t) on **x**(t + 1) is mediated by the matrix **G**(t), in a way similar to the way **F**(t) mediates the influence of **x**(t) on **x**(t + 1). Therefore, we have:

$$\boldsymbol{x}(t+1)=\boldsymbol{F}(t)\boldsymbol{x}(t)+\mathbf{G}(t)\boldsymbol{u}(t) \tag{6.1}$$

But, while the model in Equation 6.1 is suspected to be imperfect, we (fortunately) are also able to make some measurements, **z**(t), at every discrete-time iteration, which are related to the state of the system. In fact, we know how the values of the measured variables in the vector **z**(t) relate to the values in the state vector, **x**(t), at the same time:

$$\boldsymbol{z}(t)=\mathbf{H}(t)\boldsymbol{x}(t) \tag{6.2}$$

The overall goal of each Kalman Filtering iteration is to find an estimate of the state variables for the current discrete-time iteration that takes advantage of the predicted value yielded by the model (Equation 6.1), but also uses the information that can be derived from having the measurements, **z**(t), available.

So, we will focus now on the study of what happens when:

1. We can obtain the "current" state vector, from the previous state vector (through Equation 6.1),

and

2. We have just received a fresh set of measurements, **z**(t).

Notice that, accordingly, within this scope of our analysis, the symbol **x**(t) in Equation 6.1 actually represents a vector of values that were *already* obtained by the Kalman Filtering algorithm *in the previous iteration.*

6.2 THE PREDICTION PHASE

From the previous paragraphs, within our interval of analysis, **x**(t + 1) can be obtained now, through the use of Equation 6.1 (which also requires our knowledge of the **u**(t) vector). Because **x**(t + 1) is obtained from Equation 6.1, it is said that this is the "predicted" state vector, yielded by the model.

However, this "prediction" step must also provide an updated characterization of the uncertainty associated with the newly predicted state vector. Since the covariance matrix associated with **x**(t) was **P**(t), and we are obtaining **x**(t + 1) by processing **x**(t) through a linear transformation (encoded in **F**(t)), we can recall from Chapter 2 that the effect of that linear transformation on the covariance matrix of the input to the transformation will be specified by pre-multiplying the covariance by the transformation matrix and post-multiplying it by the transpose of the transformation matrix. However, Equation 6.1 also includes the presence of **u**(t), which has its uncertainty encoded in the covariance matrix **Q**(t). That uncertainty will also impact the predicted state vector. Therefore, the "prediction" equation for the covariance matrix of the state vector is:

$$\mathbf{P}(t+1) = \mathbf{F}(\mathrm{t})\mathbf{P}(\mathrm{t})\mathbf{F}(\mathrm{t})^{\mathrm{T}} + \mathbf{Q}(\mathrm{t}) \tag{6.3}$$

Equations 6.1 and 6.3 encode the first phase of each single iteration of the Kalman Filter, usually called the "Prediction Phase." In other words, by evaluating these two equations we have used the model to provide an "initial" ("draft") estimation of the state vector and its associated covariance matrix. To clarify, specifically, that the obtained values of the state vector and its covariance matrix are the result of the "Prediction Phase," based on the MODEL, we are going to re-write Equations 6.1 and 6.3 as the "prediction equations" of the Kalman Filtering algorithm, and we will include a subindex "M" for the results:

$$\boldsymbol{x}_M(t+1) = \mathbf{F}(t)\boldsymbol{x}(t) + \mathbf{G}(t)\boldsymbol{u}(t) \tag{6.4}$$

$$\mathbf{P}_M(\mathrm{t}+1) = \mathbf{F}(\mathrm{t})\mathbf{P}(\mathrm{t})\mathbf{F}(\mathrm{t})^{\mathrm{T}} + \mathbf{Q}(\mathrm{t}) \tag{6.5}$$

If we only had the model (i.e., if we were not getting new measurements z(t) with every iteration of the Kalman Filtering algorithm), we would need to stop the calculations here and would need to be content with whatever accuracy the model has in describing the evolution of the state of the system.

It is worth stopping here for a moment to realize how limiting that would be. If (as we commonly suspect in real cases) the model is not "perfect," we would still have to accept the sequence of subsequent state vectors (and their associated covariance matrices) that it produces. Even if we noticed that the variances inside the covariance matrix progressively grew (hinting that the reliability of the model was deteriorating) we would have no recourse to "check" if, indeed, the model is giving us more and more inaccurate results, and, of course, "we could not do anything about it!"

6.3 MEASUREMENTS PROVIDE A SECOND SOURCE OF KNOWLEDGE FOR STATE ESTIMATION

Fortunately, we are studying a scenario in which, in addition to our knowledge of the model equation and the initial value of the state vector (and its associated initial uncertainty, represented by the initial **P**(t)), we assume that we receive fresh values of the measurements, **z**(t), every iteration. So, for the iteration we are analyzing, we have a set of current measurement values in the vector **z**(t). Because we know that the measurements in **z**(t) relate to the state vector variables as described in Equation 6.2, we could now be tempted to adopt a completely opposite approach to the problem of estimating the value of the state variables and say . . . "Let's just get the current measurements in **z(t)** and try to deduce the values of the state variables, **x**(t), exclusively from these measurements" (from Equation 6.2). However, taking this "exclusively empirical" approach to the estimation of **x**(t) would *also* have its own drawbacks. Arguably, every measuring instrument has to acknowledge a level of uncertainty regarding the readings (measurements) that it generates. In our case, we should be able to represent the uncertainty of the measurements **z**(t) through the covariance matrix of this vector, **R**(t).

At first, we seem to be caught in a dilemma: Should we "trust" the state vector (and its covariance matrix) as provided by the model in the "Prediction Phase" equations (Equations 6.4 and 6.5), *or* should we "trust" the readings, **z**(t) (with their uncertainty, as represented by the covariance matrix, **R**(t)), from our measurement system and use those to try to generate our estimate of **x**(t)?

However, looking at the possible solution of our estimation challenge as a "binary choice" (one solution on its own *or* the other solution on its own) would be shortsighted. In a large number of circumstances the knowledge that allowed us to postulate the *model* is, at least partly, independent from the knowledge delivered to us by the *measurements* we read as we observe the actual behavior of the system. Further, the sources of uncertainty that affect the generation of predicted values by the model are likely to be (at least partially) different from the sources of uncertainty that affect the readings of our measuring equipment. Therefore, couldn't we have "the best of both worlds," drawing information from both sources of knowledge about the current state of the system? In other words, couldn't we use the current measurements from our instruments to "*enrich*" (improve) the initial state estimation that the model has provided to us in $\mathbf{x}_M(t+1)$ (and the corresponding assessment of its uncertainty, $\mathbf{P}_M(t+1)$)?

6.4 ENRICHING THE ESTIMATE THROUGH BAYESIAN ESTIMATION IN THE "CORRECTION PHASE"

Indeed, the Kalman Filtering algorithm can be understood as a process by which we seek to achieve that "enriched" estimation, in the form of a "posterior" probability density function, using the Bayesian Estimation principles we reviewed in Chapter 3. Specifically, we will use the $\mathbf{x}_M$ vector and the $\mathbf{P}_M$ covariance matrix yielded by the model through the "prediction equations" to represent the prior estimation, and the vector of measurements $\mathbf{z}(t)$ (along with its covariance matrix, $\mathbf{R}(t)$) to play the role of the added information brought in, namely the Bayesian "likelihood."

Before we proceed with the explanation of how the use of Bayesian Estimation will lead us to the last set of equations that must be implemented in each iteration of the Kalman Filtering algorithm, we want to make here a "clarifying change of variables." We introduce the two new variables $\mathbf{x}_B$ and $\mathbf{P}_B$, where the "B" subindex is meant to indicate "BEFORE," as these variables will be representing the "*prior*" probability distribution for this second "phase" of the algorithm. The change of variables is:

$$x_B = x_M(t+1) \tag{6.6}$$

$$P_B = P_M(t+1) \tag{6.7}$$

Notice that the "t + 1" on the right side of both Equations 6.6 and 6.7 implies that *the results* of Equations 6.4 and 6.5 are being copied into the new variables we will use from this point onwards (until the end of the single Kalman Filtering algorithm iteration we are studying). Back in those prediction equations the "t + 1" indices helped us understand that we were "advancing" the state of the system by one time step, for the purpose of prediction based on the model. But that part of the process is completed at this point of the execution of the algorithm. Therefore, we will take those results and use them to represent the already established or "prior" probability density function of the state estimate *in the current use of the Bayesian Estimation concept.* Furthermore, since the rest of the single iteration of the Kalman Filter we are studying takes place "in this sampling instant," the left-hand terms of Equations 6.6 and 6.7 (and all the variables we will involve in the discussions that follow) do not make explicit reference to time. *They are all "current values" of these variables.*

Another two important considerations we need to make before proceeding to the analysis of the application of Bayesian Estimation principles to the last phase of the Kalman Filtering algorithm are:

1. The (original) Kalman Filtering algorithm makes the assumption that the probability distributions involved in the analysis of the dynamic system that we are studying are all Gaussian (also called "Normal").

and

2. We need to recognize that we are, in a way, dealing with two probability spaces:
   a. The probability space defined by the variables in the state variable vector (For example, if there are only two state variables, $\mathbf{x}(t) = [x_1(t), x_2(t)]^T$.) For short, we will call this the "x space."
   b. The probability space defined by the variables in the measurements vector (For example, if there are only two measurements, $\mathbf{z}(t) = [z_1(t), z_2(t)]^T$). For short, we will call this the "z space."

Of course, the linear transformation encoded in $\mathbf{H}(t)$ enables us to map x space into z space, according to Equation 6.2, which reads $\mathbf{z}(t) = \mathbf{H}(t)\,\mathbf{x}(t)$.

Graphically, we could visualize both domains as shown in Figure 6.1 (where we are assuming Gaussian distributions involving only two state variables and two measurements, in order to be able to provide a simple visualization). It should be noticed, however, that all the explanations that follow apply to *any* number of state variables *and any* number of measurements, for as long as the distributions are Gaussian.

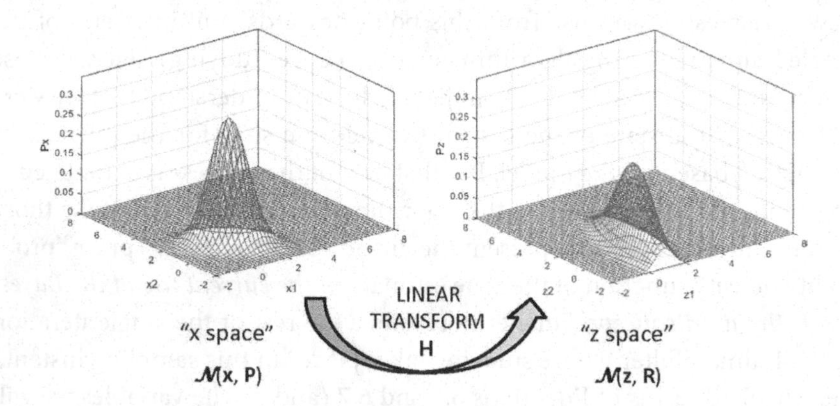

FIGURE 6.1 Illustration of the "x space" and "z space" and their relationship through the matrix H.

As indicated in Figure 6.1, the Gaussian distributions in x space and z space can be indicated as $\mathcal{N}(\mathbf{x}, \mathbf{P})$ and $\mathcal{N}(\mathbf{z}, \mathbf{R})$, respectively.

However, the Bayesian Estimation approach yields an enriched posterior distribution by multiplying a prior distribution and a Likelihood Function *in the same probability space.* Therefore, we will first map the Gaussian distribution yielded by the Prediction Phase $\mathcal{N}(\mathbf{x}_B, \mathbf{P}_B)$ to z space.

Accordingly, the Gaussian distributions we multiply to obtain the posterior distribution (in z space) are:

- Prior distribution:

$$\mathcal{N}(\mathbf{H}\mathbf{x}_B, \mathbf{H}\mathbf{P}_B\mathbf{H}^T) \tag{6.8}$$

(where we are using the conclusions reached in Chapter 2 about the mean vector and the covariance matrix of a multivariate Gaussian distribution when it is processed by a linear transformation, i.e., Equations 2.11 and 2.12)

and

- Likelihood function:

$$\mathcal{N}(\mathbf{z}, \mathbf{R}) \tag{6.9}$$

Next we recall what we learned in Chapter 2, regarding the result that is obtained when a multidimensional Gaussian function is multiplied by another multidimensional Gaussian function (Section 2.5). The result is a third multidimensional Gaussian function:

$$\text{Gaussian 0: } \mathcal{N}(\boldsymbol{\mu}_0, \boldsymbol{\Sigma}_0) \tag{6.10}$$

$$\text{Gaussian 1: } \mathcal{N}(\boldsymbol{\mu}_1, \boldsymbol{\Sigma}_1) \tag{6.11}$$

$$\text{Resulting Gaussian (Product): } \mathcal{N}(\boldsymbol{\mu}_p, \boldsymbol{\Sigma}_p) \tag{6.12}$$

where the new mean is

$$\boldsymbol{\mu}_p = \boldsymbol{\mu}_0 + \boldsymbol{K}(\boldsymbol{\mu}_1 - \boldsymbol{\mu}_0) \tag{6.13}$$

and the new covariance matrix is:

$$\boldsymbol{\Sigma}_p = \boldsymbol{\Sigma}_0 - \boldsymbol{K}\boldsymbol{\Sigma}_0 \tag{6.14}$$

In these expressions **K** is a matrix that is calculated from the two initial covariance matrices:

$$K = \Sigma_0 \left(\Sigma_0 + \Sigma_1\right)^{-1} \tag{6.15}$$

Therefore, the "enriched" posterior distribution, in z space, will be a Gaussian distribution with:

Mean vector:

$$Hx_B + K\left(z - Hx_B\right) \tag{6.17}$$

and covariance matrix:

$$HP_BH^T - KHP_BH^T \tag{6.18}$$

Where

$$K = HP_BH^T\left(HP_BH^T + R\right)^{-1} \tag{6.19}$$

according to Equation 6.15.

After obtaining the results for the posterior distribution in z space indicated in Expressions 6.17 and 6.18 (with the definition for K given by Equation 6.19), we need to stop and realize that what we are pursuing is the knowledge of the distribution *in x space* that corresponds to the improved posterior distribution that we just defined in z space. Let's assign to that distribution in x space that will correspond to the posterior distribution in z space the variables $\mathbf{x}_A$ and $\mathbf{P}_A$ for the mean vector and the corresponding covariance matrix. That is, the posterior distribution in x space that we are seeking is:

$$\mathcal{N}\left(x_A, P_A\right) \tag{6.20}$$

It should be noticed that the ultimate goal of each iteration of the Kalman Filtering algorithm is precisely the calculation of these "enriched" posterior distribution parameters (in x space), $\mathbf{x}_A$ and $\mathbf{P}_A$. (We use the A subscript to mean "AFTER," for posterior distribution).

Since $\mathbf{x}_A$ and $\mathbf{P}_A$ would be mapped to z space through the linear transformation represented by **H**, those (to be determined) parameters define a Gaussian distribution in z space with the following:

- Mean vector:

$$Hx_A \tag{6.21}$$

- Covariance matrix:

$$HP_AH^T \tag{6.22}$$

Notice that we have arrived at expressions for the mean vector and the covariance matrix of the posterior distribution in z space from two "routes": From the application of Bayesian Estimation in z space (Expressions 6.17 and 6.18), and from the mapping of the desired x space parameters, $\mathbf{x_A}$ and $\mathbf{P_A}$, to the z space (Expressions 6.21 and 6.22). However, ultimately both pairs of equations refer to the same entities (e.g., the mean vectors described by Expression 6.17 and by Expression 6.21 are actually the same). Therefore, we can establish the following two equalities:

$$Hx_A = Hx_B + K(z - Hx_B) \tag{6.23}$$

and

$$HP_AH^T = HP_BH^T - KHP_BH^T \tag{6.24}$$

We introduce a *new* matrix $\mathbf{K_G}$:

$$K_G = P_BH^T\left(HP_BH^T + R\right)^{-1} \tag{6.25}$$

such that we can substitute every occurrence of **K** with the product **H** $\mathbf{K_G}$ (compare Equations 6.19 and 6.25) in Equations 6.23 and 6.24:

$$Hx_A = Hx_B + HK_G(z - Hx_B) \tag{6.26}$$

and

$$HP_AH^T = HP_BH^T - HK_GHP_BH^T \tag{6.27}$$

From Equations 6.26 and 6.27, we have:

$$Hx_A = H\{x_B + K_G(z - Hx_B)\} \tag{6.28}$$

and

$$HP_AH^T = H\{P_B - K_GHP_B\}H^T \tag{6.29}$$

Looking at Equation 6.28, we notice that on both sides of the equality there are products where the first factor is **H**. The equality will be verified if the second factors on both sides of Equation 6.28 are the same. Therefore, a solution for $\mathbf{x}_A$ is:

$$\boldsymbol{x}_A = \boldsymbol{x}_B + \boldsymbol{K}_G\left(\boldsymbol{z} - \boldsymbol{H}\boldsymbol{x}_B\right) \tag{6.30}$$

Similarly, focusing on the term that appears as a factor between **H** and $\mathbf{H}^T$ on both sides of Equation 6.29, a solution for $\mathbf{P}_A$ is:

$$\boldsymbol{P}_A = \boldsymbol{P}_B - \boldsymbol{K}_G\boldsymbol{H}\boldsymbol{P}_B \tag{6.31}$$

Equations 6.30 and 6.31 are the "correction equations" (or "update equations") for the Kalman Filtering algorithm, where the "Kalman Gain," $\mathbf{K}_G$ (repeated from Eq. 6.25), is:

$$\boldsymbol{K}_G = \boldsymbol{P}_B\boldsymbol{H}^T\left(\boldsymbol{H}\boldsymbol{P}_B\boldsymbol{H}^T + \boldsymbol{R}\right)^{-1} \tag{6.32}$$

The correction equations allow us to obtain the posterior $\mathbf{x}_A$ and $\mathbf{P}_A$, which have the full benefit of the Bayesian Estimation for this iteration and will be fed back as the inputs **x**(t) and **P**(t) for the prediction step of the *next* iteration.

Therefore, as promised at the beginning of this chapter, this is the ALGORITHM for (one iteration of) the DISCRETE KALMAN FILTER:

Prediction equations:

$$\boxed{\boldsymbol{x}_M(t+1) = \mathbf{F}(t)\boldsymbol{x}(t) + \mathbf{G}(t)\boldsymbol{u}(t)} \tag{6.33}$$

$$\boxed{\mathbf{P}_M(t+1) = \mathbf{F}(t)\mathbf{P}(t)\mathbf{F}(t)^T + \mathbf{Q}(t)} \tag{6.34}$$

Change of variables—only for clarification:

$$\boxed{\boldsymbol{x}_B = \boldsymbol{x}_M(t+1)} \tag{6.35}$$

$$\boxed{\boldsymbol{P}_B = \boldsymbol{P}_M(t+1)} \tag{6.36}$$

Correction equations" or "update equations:

$$\boxed{\boldsymbol{K}_G = \boldsymbol{P}_B\boldsymbol{H}^T\left(\boldsymbol{H}\boldsymbol{P}_B\boldsymbol{H}^T + \boldsymbol{R}\right)^{-1}} \tag{6.37}$$

$$\boxed{x_A = x_B + K_G\left(z - Hx_B\right)} \tag{6.38}$$

$$\boxed{P_A = P_B - K_G H P_B} \tag{6.39}$$

We want to emphasize, again, that the "Change of variables—only for clarification" specified by Equations 6.35 and 6.36 *does not involve any computation* and that these equations are inserted only to help the reader differentiate the extent and the reasoning behind the two different phases of the Kalman Filter: "Prediction" and "Correction (or Update)." Many other sources will, indeed, show the Kalman Filtering algorithm as comprised of only five equations, where the variables are "not changed" from the prediction to the correction equations. Further, it is true that the algorithm can be implemented in just five lines of MATLAB® code, as opposed to the seven lines that we will use in the MATLAB® function we develop in Chapter 8. We have sacrificed compactness in the algorithm presentation in an effort to enhance clarity.

CHAPTER 7

Reflecting on the Meaning and Evolution of the Entities in the Kalman Filter Algorithm

We have explained and stated the computational steps involved in the Discrete Kalman Filter algorithm (Equations 6.33 to 6.39). We must now stop to reflect on what is the expected benefit of the algorithm and the interpretation of its variables and parameters, as the algorithm is applied iteratively. This will be a good preparation for the development of a block diagram of the algorithm and a MATLAB® implementation for it, in the next chapter.

7.1 SO, WHAT IS THE KALMAN FILTER EXPECTED TO ACHIEVE?

The Kalman Filter is a mechanism to obtain a particularly good estimation of the state vector, in $\mathbf{x}_A$, from two sources of information: the sequence of recurrent model results, $\mathbf{x}_B = \mathbf{x}_M(t + 1)$, and the information brought in by the instrumental measurements, $\mathbf{z}(t)$, that we receive for the execution of every iteration of the algorithm. In fact, it has been demonstrated to be an "*optimal*" estimator for this scenario, under specific conditions (e.g., all the distributions involved being Gaussian). In particular, "it can be shown that of all possible filters, it is the one that *minimizes* the variance of the estimation error" (Simon 2001). What does that mean for us, practically? It means that the algorithm will strive to provide us in its output, $\mathbf{x}_A$, state estimates that have less uncertainty associated to them than the state estimates that are produced by the model alone, $\mathbf{x}_B = \mathbf{x}_M(t + 1)$, and less uncertainty that the estimates we could try to derive exclusively from the empirical measurements of system performance, $\mathbf{z}(t)$.

This expectation is reasonable, as we have seen that the Kalman Filter is "enriching" the state output from the model, used as "prior" information, with the "likelihood" provided by the measurements acquired, to yield an improved "posterior" state estimation through the application of Bayesian Estimation principles.

The external uncertainty affecting the predictions of the model and the uncertainty of the measurements, encoded by the covariance matrices $\mathbf{Q}$ and $\mathbf{R}$, may be constant, and determined by the definition of the model and the accuracy ratings of the instruments used to obtain measurements. On the other hand, the recursive computation of the state covariance matrix output, in each iteration of the Kalman Filter algorithm, provides us with a way to monitor how the uncertainty of our effective estimate, $\mathbf{x}_A$, is evolving. The expectation is that, as the Kalman Filter process evolves through iterations, the uncertainty of its estimate should be minimized, which might be reflected, for example, in the decrease of the variances contained in the main diagonal of $\mathbf{P}_A$.

7.2 EACH ITERATION OF THE KALMAN FILTER SPANS "TWO TIMES" AND "TWO SPACES"

During the development of the Kalman Filter algorithm we found that the state vector estimation is predicted from its value at time "t," to its resulting value at time "t + 1." In both cases it is accompanied by its covariance matrix, $\mathbf{P}$, quantifying its uncertainty. Therefore, it is clear that each Kalman Filter iteration involves two consecutive sampling times: "t," followed by "t + 1."

Similarly, we mentioned that there are two probability spaces at play. First, one defined by the joint distribution of the variables contained in the state variable vector, $\mathbf{x}$, and then one defined by the joint distribution of the measurements contained in vector $\mathbf{z}$. These spaces are associated according to the measurement equation, $\mathbf{z} = \mathbf{Hx}$ (which, implicitly, also associates the corresponding covariance matrices). While the application of the Bayesian Estimation process may be thought of as being applied in the "z space," the overall goal of the Kalman Filter iteration is to obtain estimates of the state variables. Therefore, the outputs of each iteration, namely $\mathbf{x}_A$ and $\mathbf{P}_A$, are referred to "x space."

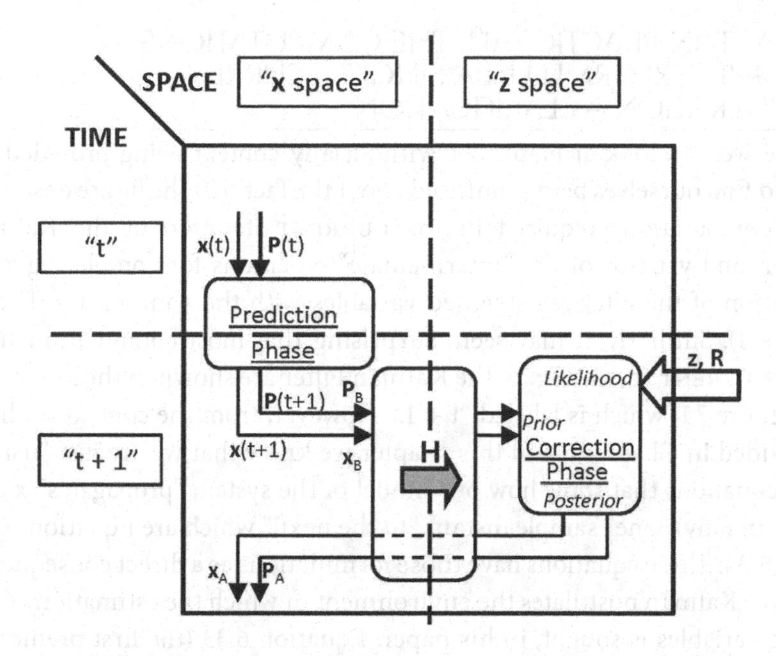

FIGURE 7.1 Depiction of the two "times" and two "spaces" considered in each iteration of the Kalman Filter, identifying the phases of the process and the variables of the algorithm.

Figure 7.1 attempts to illustrate both times and both spaces, and places the different estimate and measurement vectors and their corresponding covariance matrices in the resulting global context.

The *Prediction Phase* advances the model estimate **x**(t) to the one for the next time step, **x**(t + 1), and in doing so it also advances the associated covariance matrix from **P**(t) to **P**(t + 1). In this book, we have chosen to implement a change of variable, calling these results $\mathbf{x_B}$ and $\mathbf{P_B}$, dropping the reference to one of the time instances or the other, because all the remaining computations will be considered to take place "at the current time" (which is "t + 1" in the figure). Both $\mathbf{x_B}$ and $\mathbf{P_B}$ are projected from "x space" to "z space," according to the measurement equation, $\mathbf{z} = \mathbf{Hx}$, so that, once projected, they can play the role of the prior distribution for the Bayesian Estimation process. The likelihood role is implemented by the insertion of new (measurement) data in vector **z**, whose covariance matrix is **R**. The Bayesian Estimation process in the *Correction Phase* yields a posterior distribution. However, the equations developed in Chapter 6 are manipulated to yield the "x-space" counterparts of the posterior distribution in "z-space." These are $\mathbf{x_A}$ and $\mathbf{P_A}$, the final results of the complete Kalman Filter iteration.

7.3 YET, IN PRACTICE ALL THE COMPUTATIONS ARE PERFORMED IN A SINGLE, "CURRENT" ITERATION—CLARIFICATION

If we were to look at Figure 7.1 without any context being provided we could find ourselves being confused about the fact that the figure describes the computations required in the CURRENT iteration of the Kalman Filter, and yet two of the "intermediate" equations for completing THIS iteration of the filter are assigned variables with the names $\mathbf{x}(t + 1)$ and $\mathbf{P}(t + 1)$. Similarly, it may seem surprising that most computations FOR THE CURRENT ITERATION of the Kalman Filter are shown in the lower half of Figure 7.1, which is labeled "t + 1." However, from the context we have provided in Chapter 6 and this chapter we know that we are just "using" the equations that show how our model of the system "propagates" $\mathbf{x}$ and $\mathbf{P}$ from (any) "one" sample instant "to the next," which are Equations 6.33 and 6.34. These equations have those formulations as a direct consequence of how Kalman postulates the environment in which the estimation of the state variables is sought, in his paper. Equation 6.33 (the first prediction equation, i.e., the "model equation") can be read directly from Figure 2 in Kalman's paper (Kalman 1960), which we reproduced in this book as Figure 5.1, and appears as Equation (16) in his paper (although using different variable names, which we detailed in Chapter 5). As we started the study of the state estimation process via Kalman Filtering from that figure, we have resolved to keep that format of the "model equation," showing $\mathbf{x}(t + 1)$ as a result, although we are calculating those values and using them as input to the Correction Phase in THE CURRENT iteration. The same considerations apply to the second prediction equation, Equation 6.34, which shows $\mathbf{P}(t + 1)$ as result.

We have attempted to avoid confusion by dropping any reference to time in the remainder of the processing needed to complete the current iteration of the Kalman Filter. That is why we introduced Equations 6.35 and 6.36, transferring the result of the prediction equations to variables that are simply characterized as the prior estimates for THE CURRENT Bayesian Estimation process, taking place in the Correction Phase of the current Kalman Filter iteration: $\mathbf{x}_B = \mathbf{x}_M(t + 1)$; $\mathbf{P}_B = \mathbf{P}_M(t + 1)$.

The reader might find that other texts, for example (Bishop and Welch 2001), present prediction equations that make it more explicit that in the Prediction Phase we are using values previously created ($\mathbf{x}_{k-1}$ and $\mathbf{P}_{k-1}$) to obtain "current" values ($\mathbf{x}_k$ and $\mathbf{P}_k$) in this iteration. (Those equations read, approximately: $\mathbf{x}_k = \mathbf{A}\mathbf{x}_{k-1} + \mathbf{B}\mathbf{u}_{k-1}$, and $\mathbf{P}_k = \mathbf{A}\mathbf{P}_{k-1}\mathbf{A}^T + \mathbf{Q}$). However, as explained earlier, we resolved to keep the original formulation of the model equation.

We believe that this aspect will be further clarified in Chapter 8, when we develop the MATLAB® function to implement one (the "current") iteration of the Kalman Filter algorithm and show a figure representing how that function is called iteratively (Figure 8.2 in Chapter 8).

7.4 MODEL OR MEASUREMENT? K_G DECIDES WHO WE SHOULD TRUST

In the previous chapter, when we were trying to develop the Kalman Filtering algorithm we first seemed to be caught in a dilemma:

> Should we "trust" the state vector (and its covariance matrix) as provided by the model in the "Prediction Phase" Equations (Equations 6.4 and 6.5), *or* should we "trust" the readings, $\mathbf{z}(t)$ (with their uncertainty, as represented by the covariance matrix, $\mathbf{R}(t)$), from our measurement system and use those to try to generate our estimate of $\mathbf{x}(t)$?

We later found that the Kalman Filter algorithm actually *combines the information from both sources* to yield an *improved estimate*. However, the question remains: In which proportion is our final estimate, $\mathcal{N}(\mathbf{x_A}, \mathbf{P_A})$, based on the information provided by the model (mapped to z space), $\mathcal{N}(\mathbf{Hx_B}, \mathbf{HP_BH^T})$?, and on which proportion is it based on the information provided by the measurements, $\mathcal{N}(\mathbf{z}, \mathbf{R})$?

The Kalman Filter does not give a fixed answer to the question posed at the beginning of this section. Instead, it will create a new assessment of the levels of involvement for the model and the measurements in a dynamic way, *optimizing this decision for each iteration.* This is a consequence of the fact that it is executing a Bayesian Estimation process in the Correction Phase of each iteration. Bayesian Estimation obtains the "posterior" Gaussian distribution as *the product* of the Gaussian likelihood (in this case from the measurements), and the "prior" Gaussian distribution (in this case the result of one-time-step update in the model).

We will first examine this process in the simplest possible case, in which there would be *just one state variable*, which we will represent in this discussion with the scalar x, *and just one* measurement, which we will represent with the scalar z. Under those conditions $x = hz$, where h is just a scalar value (instead of the matrix $\mathbf{H}$, in the general case).

In this case all the probability functions are univariate, and we can apply to the product of the likelihood times the prior density what we learned in Chapter 1 (Section 1.5). That is, the product of two Gaussian distributions $\mathcal{N}(\mu_1, \sigma^2_1)$ and $\mathcal{N}(\mu_2, \sigma^2_2)$ yields the following mean and variance for the resulting Gaussian distribution $\mathcal{N}(\mu_p, \sigma^2_p)$:

$$\mu_p = \mu_2 + \frac{\sigma_2^2(\mu_1 - \mu_2)}{\sigma_1^2 + \sigma_2^2} \tag{7.1}$$

and

$$\sigma_p^2 = \sigma_2^2 - \frac{\sigma_2^4}{\sigma_1^2 + \sigma_2^2} \tag{7.2}$$

And, if we define the parameter k (which is a scalar, for the univariate case) as this ratio of variances:

$$k = \frac{\sigma_2^2}{\sigma_1^2 + \sigma_2^2} \tag{7.3}$$

Then

$$\mu_p = \mu_2 + k(\mu_1 - \mu_2) \tag{7.4}$$

and

$$\sigma_p^2 = \sigma_2^2 - k\sigma_2^2 = \sigma_2^2(1 - k) \tag{7.5}$$

In the specific univariate case we are now studying, the two Gaussian functions being multiplied in the Bayesian Estimation process will be the one for the measurement, z, that is, $\mathcal{N}(\mu_z, \sigma^2_z)$ and the Gaussian that results from projecting the single state variable from the *model* to the z space, which we will denote by "m," that is $\mathcal{N}(\mu_m, \sigma^2_m)$. (This mapping would be according to the scalar equation m = hx.)

We will use the results from Chapter 1 with these substitutions:

"Gaussian 1": Gaussian from measurements: $\mathcal{N}(\mu_z, \sigma^2_z)$.

"Gaussian 2": Gaussian from the model results, projected to z space: $\mathcal{N}(\mu_m, \sigma^2_m)$.

With these substitutions, the resulting mean (of the product) will be:

$$\mu_P = \mu_m + k(\mu_z - \mu_m) \tag{7.6}$$

And the resulting variance (of the product) will be:

$$\sigma_p^2 = \sigma_m^2 - k\sigma_m^2 = \sigma_m^2(1-k) \tag{7.7}$$

with the scalar k defined in this case as:

$$k = \frac{\sigma_m^2}{\sigma_m^2 + \sigma_z^2} \tag{7.8}$$

Let's now explore two "extreme" cases in these two equations and determine their effects. These "extreme" cases will address the instances when one of the sources of information (e.g., the model) has a very low level of uncertainty, while the other source of information (e.g., the measurements) has a much higher level of uncertainty (i.e., $\sigma^2_m << \sigma^2_z$), and vice versa.

But first let's consider the meaning and range of values for k, according to Equation 7.8. Since both variances are positive real numbers, k can range from values close to zero, when $\sigma^2_z >> \sigma^2_m$, to values close to 1, when $\sigma^2_z << \sigma^2_m$. That is, k represents, in a scale approximately from 0 to 1, how much we should pay preferential attention to the information from the measurement, over the information from the model (because the uncertainty of the measurement is, comparatively, smaller). With this in mind, we can interpret Equation 7.6 as the mechanism in which the Kalman Filter will "correct" or "override" the resulting mean, μ_p, from the "default" prior value proposed by the model, μ_m, to the alternative value suggested by the measurement, μ_z.

Notice that in Equation 7.6, if k were to take the value of ~0 (implying that the uncertainty in the measurement is much higher than the uncertainty in the model, $\sigma^2_z >> \sigma^2_m$), then Equation 7.6 would implement minimal correction, yielding $\mu_p \sim \mu_m$. This would be an advantageous choice, given the circumstances.

In contrast, if k were to take the value of ~1 (implying that the uncertainty in the model is much higher than the uncertainty in the measurement, $\sigma^2_m >> \sigma^2_z$), then $\mu_p \sim \mu_z$, which is also the right choice, for this second scenario! That is, in this case (k ~ 1), the Kalman Filter would be applying (almost) the full "correction" that is suggested in the parenthesis ($\mu_z - \mu_m$) of Equation 7.6 to obtain a final result that is very close to μ_z. This is one of the reasons why we have preferred to call the second phase of the Kalman Filter algorithm the "Correction Phase," although it is called the "Update Phase" in other texts.

Similarly, we can analyze what happens in these extreme cases with respect to the variance of the result, σ^2_p, observing Equation 7.7. Particularly, when we focus on the format: $\sigma^2_p = \sigma^2_m (1 - k)$, we realize that the variance of the result will be a fraction of the variance of the model, since the factor in parenthesis, i.e., $(1 - k)$, will be less than 1. This is already good news, because it means that the resulting estimate will have less uncertainty than the estimate generated by the model alone. But it is still very interesting to see what happens in each of the two "extreme cases" we are studying.

When the uncertainty of the measurement is much bigger than the uncertainty of the model, $\sigma^2_z >> \sigma^2_m$, then k will be close to 0 and, therefore, the variance of the result will be close to that of the model, which was small (by comparison), to begin with.

When the uncertainty of the model is much bigger than the uncertainty of the measurement, ($\sigma^2_m >> \sigma^2_z$), then k will be close to 1 and, therefore, the variance of the result will be just a small fraction $(1 - k)$ of the original (large) variance of the model. This is also a beneficial result.

Fortunately, then, in both scenarios we have discussed, the variance of the product ends up being (relatively) small, reflecting that we are obtaining a result with (comparatively) reduced uncertainty. This confirms the benefits of the merger of information from the model and the measurements performed by the Kalman Filter.

The shifting of the mean and the reduction in variance that we observe in the resulting Gaussian is an inherent effect of the product of two Gaussians. If one of the two Gaussian factors is particularly "slim," its values will decay significantly and quickly for values of the random variable away from the corresponding mean. Those low values will strongly reduce the impact of the other Gaussian in the product for values of the random variable in the ranges where the other Gaussian had its highest values. The net effects are a "shifting" and a "slimming," so that the resulting Gaussian is more similar to the "slim" Gaussian factor than to the "broad" Gaussian factor, with respect to both placement and shape. This can be seen in the following MATLAB® example in which the original Gaussians were simulated as $\mathcal{N}(\mu_m, \sigma^2_m) = \mathcal{N}(1, 1)$ (shown with dashed lines in Figure 7.2) and

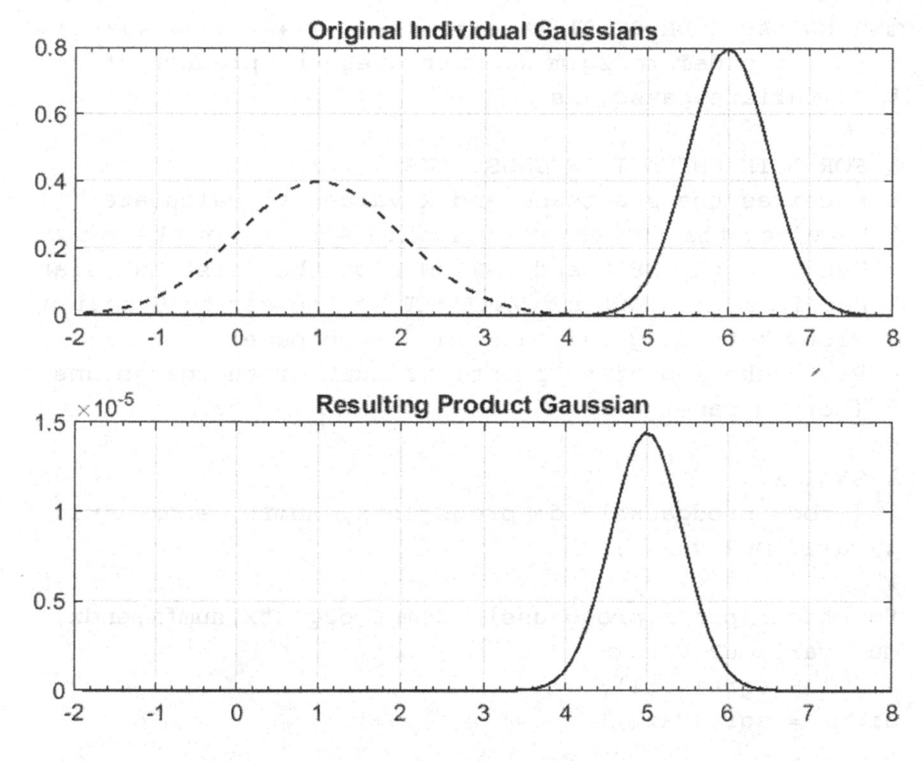

FIGURE 7.2 Demonstration of the product of two univariate Gaussian functions, simulated with the MATLAB® function [prodx, prodgauss] = demprod2g(−2, 100, 8, 1, 1, 6, 0.25).

$\mathcal{N}(\mu_z, \sigma^2_z) = \mathcal{N}(6, 0.25)$ (shown with solid lines in the figure). With those variances, $k = 1/(0.25 + 1) = 0.8$. The shape of the Gaussian that results by calculating the (point-to-point) product of the two original Gaussians confirms that, in agreement with Equations 7.6 and 7.7, the mean and the variance of the resulting Gaussian are: $\mu_p = 5$ and $\sigma^2_p = 0.2$.

The simulation was obtained with the MATLAB® function demprod2g.m (listed in MATLAB® Code 07.01), which calls the function calcgauss.m (also listed, as MATLAB® Code 07.02). The invocation of the function was:

[prodx, prodgauss] = demprod2g(−2,100,8,1,1,6,0.25).

This means that the parameters simulated were: Mean for m = 1; variance for m = 1; mean for z = 6; variance for z = 0.25.

```
%%% MATLAB CODE 07.01 ++++++++++++++++++++++++++++++++
% function demprod2g.m demonstrates the product of 2
% univariate gaussians
%
% FOR BOTH UNIVARIATE GAUSSIANS
% Receives the start and end x values to calculate
% Receives the number of evaluations to do in the range
% Receives the MEAN and VARIANCE of the first Gaussian
% Receives the MEAN and VARIANCE of the second Gaussian
% Plots both original gaussians (top pane)
% Plots the (point-to-point) product of the gaussians
% (bottom pane)
%
% SYNTAX:
% [prodx,prodgauss]= demprod2g(stx, numfx, endx, mu1,
% var1, mu2,var2);
%
function [prodx,prodgauss]= demprod2g(stx,numfx,endx,
mu1,var1,mu2,var2);
sigm1 = sqrt(var1);
sigm2 = sqrt(var2);

 [valsx,vgauss1] =calcgauss(stx,numfx,endx,mu1, sigm1);
 [valsx,vgauss2] =calcgauss(stx,numfx,endx, mu2,sigm2);

prodx = valsx;
prodgauss = vgauss1 .* vgauss2;

% Plotting
figure;
subplot(2,1,1); plot(prodx,vgauss1,'b--');
hold on
plot(prodx,vgauss2,'r');
hold off
title('Original individual Gaussians');
grid
set(gca,'XMinorTick','on')
grid minor
set(gca,'Xtick',stx:1:endx)
grid off
grid on
```

```
subplot(2,1,2); plot(prodx,prodgauss,'b');
title('Resulting Product Gaussian');
grid
set(gca,'XMinorTick','on')
grid minor
set(gca,'Xtick',stx:1:endx)
grid off
grid on

end
%%% MATLAB CODE 07.01 +++++++++++++++++++++++++++++++

%%% MATLAB CODE 07.02 +++++++++++++++++++++++++++++++
% calcgauss.m
% evaluates a normal pdf from the theoretical formula,
% providing start x, number of x, end x
% and mean and std.
% RETURNS the vector of values calculated (for plot)
%
% SYNTAX: [valsx,resgauss] = calcgauss( startx, numofx,
%endx, mu, sigm);
%
%
function [valsx,resgauss] =calcgauss(startx,numofx,
endx,mu,sigm);
%
% Vector of evaluation values (abcisas)
gapx = (endx-startx)/(numofx-1);
valsx = zeros(numofx, 1);
for i = 1:numofx
          valsx(i,1) = startx + (i * gapx);
end

coef = 1/(sqrt(2 * pi * sigm^2));
dnm = 2 * sigm^2;

resgauss = coef .* exp(((-1) * (valsx-mu).^2)./ dnm);
end
%%% MATLAB CODE 07.02 +++++++++++++++++++++++++++++++
```

To reaffirm the intuitive understanding of the role that the parameter k plays in the shifting of the mean for the product of two Gaussians, Figures 7.3 and 7.4 propose an analogy of this effect with the principle of operation of the iron-vane or moving-iron rudimentary galvanometer, as presented by Holman (Holman 2001). In the galvanometer (Figure 7.3), the current to be measured, *i*, creates a magnetic field that will make the iron plunger move clockwise. The higher the current, the higher the force on the iron plunger, which will yield a higher reading in the scale, by reaching equilibrium with the force of the spring at a larger displacement angle. In the product of Gaussians (Figure 7.4), if, for example $\sigma^2_m >> \sigma^2_z$, then k will be "large" (close to 1) and it will "pull" the mean of the product Gaussian very close to μ_z, and display a variance that is close to

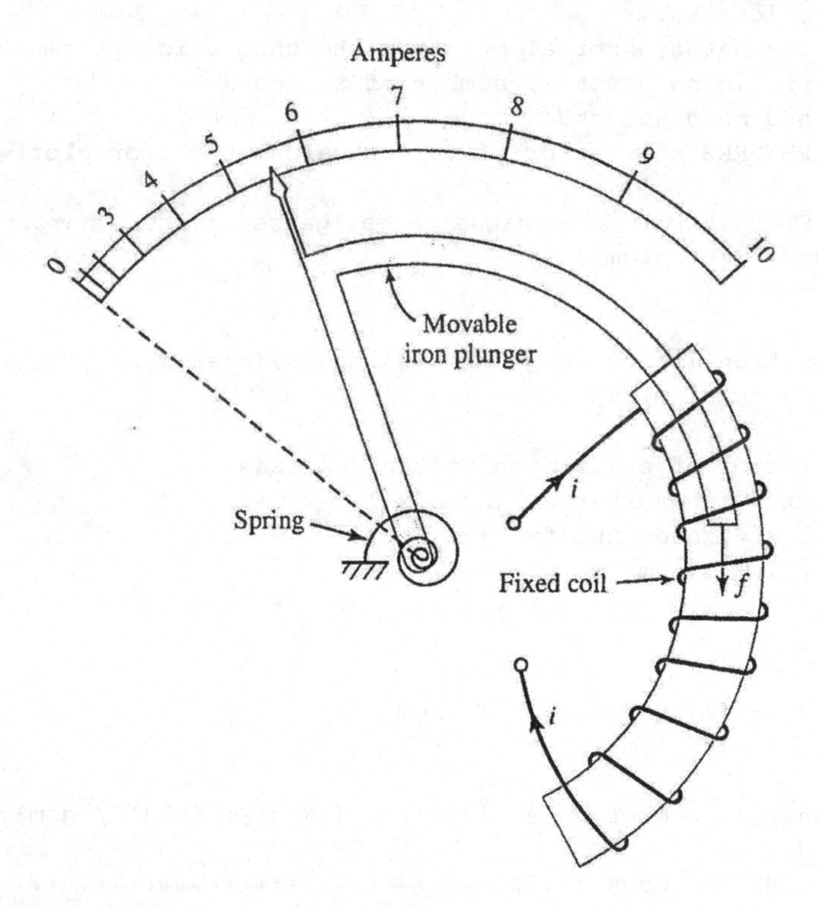

FIGURE 7.3 Principle of operation of the iron-vane or moving-iron instrument. (From Holman, 2001. Used with permission from McGraw-Hill Inc.)

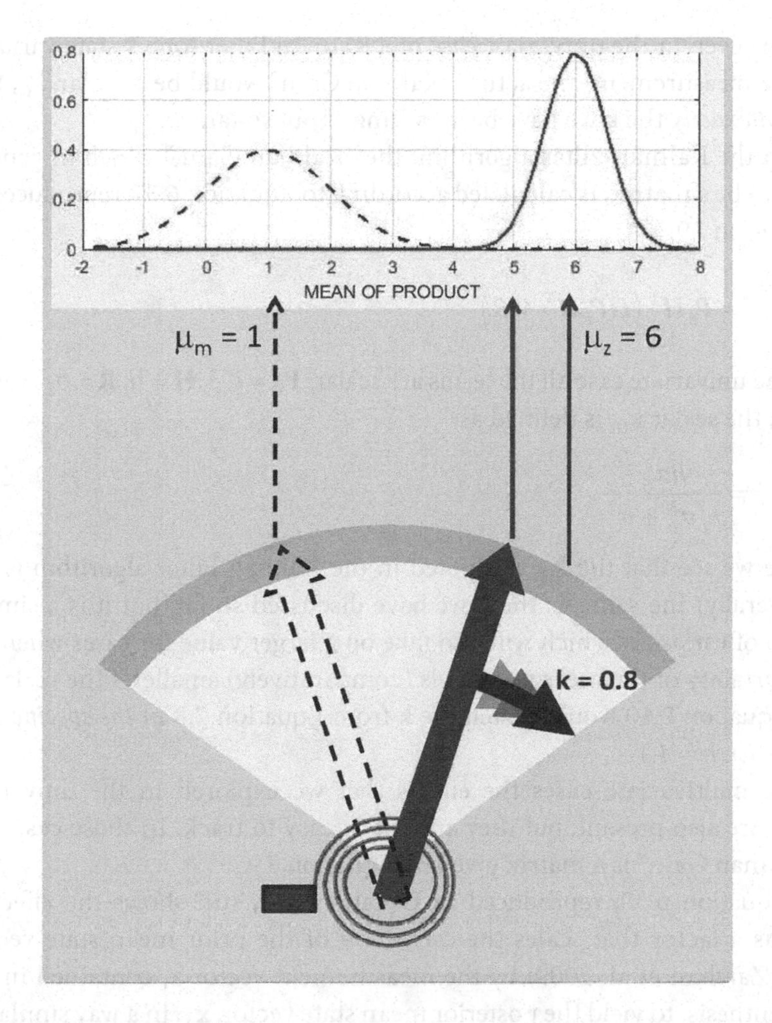

FIGURE 7.4 Analogy of mean correction in a product of Gaussians, driven by k, compared to the principle of operation of the iron-vane instrument.

the (relatively small) σ^2_z. This is what happened in the case simulated in Figure 7.2, where k = 0.8.

In this discussion we have considered the effects that take place when we obtain the product of two Gaussian functions, because that kind of product is at the heart of the Bayesian Estimation process that is implemented in the Correction Phase of the Kalman Filter iteration. Looking at a generic product of two univariate Gaussian functions has facilitated the perception and interpretation of the effect we have discussed. However, this phenomenon exists in a slightly modified form within the Kalman

Filter. Even in the univariate case of a Kalman Filter (one x state variable, one z measurement) the actual "Kalman Gain" would be a scalar, k_G, that *is not exactly* the k we have been talking about so far.

In the Kalman Filter algorithm, the "Kalman Gain," which in general would be a matrix, is calculated according to Equation 6.37, reproduced in Equation 7.9:

$$K_G = P_B H^T \left(H P_B H^T + R \right)^{-1} \tag{7.9}$$

In the univariate case all the terms are scalar: $\mathbf{P_B} = \sigma_B^2$, $\mathbf{H} = h$, $\mathbf{R} = \sigma_z^2$; therefore, the scalar k_G, is defined as:

$$k_G = \frac{h\sigma_B^2}{h^2\sigma_B^2 + \sigma_z^2} \tag{7.10}$$

Here we see that the k_G, computed in the Kalman Filter algorithm is not (generally) the same as the k we have discussed so far, but it is a similar ratio of variances which will also take on a larger value for cases when the uncertainty of the measurement is (comparatively) smaller. (The scalar k_G in Equation 7.10 would match the k from Equation 7.8 *in the specific case in which h = 1.)*

In multivariate cases the effects that we explored in the univariate case are also present, but they are not as easy to track. In those cases the "Kalman Gain" is a matrix given by Equation 7.9.

Equation 6.38, reproduced in Equation 7.11, still shows the effect of $\mathbf{K_G}$ as a factor that scales the *correction* of the prior mean state vector, $\mathbf{x_B}$, (Zarchan et al. 2009) by the measurement vector $\mathbf{z}$, contained in the parenthesis, to yield the posterior mean state vector, $\mathbf{x_A}$, in a way similar to what happened in Equation 7.6 for the univariate case we discussed.

$$x_A = x_B + K_G \left(z - H x_B \right) \tag{7.11}$$

Similarly, Equation 6.39, reproduced as Equation 7.12, shows the effect of $\mathbf{K_G}$ in the modification of the uncertainty (represented by the covariance matrix) from the prior distribution, $\mathbf{P_B}$, to the posterior distribution, $\mathbf{P_A}$. The similarity of the effect can best be perceived when we compare Equation 7.12 with Equation 7.7, for the product of two univariate Gaussians, which reads $\sigma_p^2 = \sigma_m^2 - k\,\sigma_m^2$.

$$P_A = P_B - K_G H P_B \tag{7.12}$$

III

Examples in MATLAB®

THIS PART OF THE book starts with the creation of a general MATLAB® function that implements a single iteration of the Discrete Kalman Filtering algorithm, by mapping the equations defined at the end of Chapter 6 to MATLAB® code. The function is set up in a very general format, so it can be used as the basis for complete Kalman Filtering solutions under a range of circumstances. In addition, Chapter 8 provides a block diagram for the function and an illustration of how the replication of the block diagram functionality through time yields a complete Kalman Filter implementation. Then, in Chapter 9, we use the MATLAB® function created to simulate, resolve and analyze the univariate estimation scenario first presented in Chapter 4 (inaccessible voltage estimation). In Chapter 10 we use the same MATLAB® function to simulate, resolve and analyze the "falling wad of paper" multivariate scenario described and modeled in Chapter 5. The reader is strongly encouraged to make full use of the listings we provide of all the MATLAB® code used, by changing parameters to simulate alternative scenarios and to inspect the 3-dimensional figures generated interactively, using the 3-D rotation capabilities of MATLAB® visualizations.

CHAPTER 8

MATLAB® Function to Implement and Exemplify the Kalman Filter

In this chapter we will develop a simple MATLAB® function that implements the computations necessary to complete both the "Prediction Phase" and the "Correction (or Update) Phase" of a single iteration of the Discrete Kalman Filter. We will start by re-stating the two prediction equations and the three correction equations that were found in Chapter 6, with the computation of the Kalman Gain (K_G) as the first calculation needed for the Correction Phase. Fortunately, we will see that MATLAB® allows a simple one-to-one mapping from the set of equations to this function. This is the set of calculations that needs to be executed, recursively, by a MATLAB® program or a real-time system to actually implement a Kalman Filter. The single-iteration process will be illustrated as a block diagram, and the iterative execution of that process will be illustrated by a series of copies of the one-iteration block diagram.

8.1 DATA AND COMPUTATIONS NEEDED FOR THE IMPLEMENTATION OF ONE ITERATION OF THE KALMAN FILTER

As we found out, an iteration of the Kalman Filter needs to first obtain a preliminary estimate for the state vector, for this moment in time, according to the model available for the system. The system model is set up for providing one-step predictions. Therefore, if we call the current state vector **x**(t), the estimate of the state vector predicted by the model will be **x**(t + 1). Similarly, the predicted uncertainty of the estimate, **P**(t + 1), is obtained from **P**(t). This is achieved through the prediction equations:

"PREDICTION EQUATIONS":

$$\boldsymbol{x}_M(t+1) = \boldsymbol{F}(t)\boldsymbol{x}(t) + \boldsymbol{G}(t)\boldsymbol{u}(t) \tag{8.1}$$

$$\boldsymbol{P}_M(t+1) = \boldsymbol{F}(t)\boldsymbol{P}(t)\boldsymbol{F}(t)^T + \boldsymbol{Q}(t) \tag{8.2}$$

From these equations we realize that, in addition to the availability of **x**(t) and **P**(t), which would have been obtained as results from a previous iteration of the Kalman Filter, or as initial values, we need the current values of the matrices **F**(t), **G**(t) and **Q**(t). If we are dealing with a time-varying model, we would need to know the correct values of **F**(t) and **G**(t) for "this" time. Although the formulation of the Kalman Filter allows for this possibility, it is common to deal with time-invariant systems where the matrices **F**(t) and **G**(t) in the model are constant. Similarly, although the framework allows for a changing characterization of the external uncertainty, represented by **Q**(t), it is common that we might only have access to a unique, constant characterization of this uncertainty.

If the system actually receives "control inputs" at every iteration, the current values of those signals would need to be provided in the vector **u**(t). Some systems do not receive "control inputs," and for others the values in vector **u**(t) might be constants.

As we indicated in Chapter 6, once the values of $\mathbf{x}_M$(t + 1) and $\mathbf{P}_M$(t + 1) are obtained from the model, the Prediction Phase is complete and those values will be used as "prior information" (i.e., the information we have before the enrichment of the information through combination with the current values of measurements, available in this iteration). Therefore, to make it completely clear that we continue now with the "Correction Phase" which applies Bayesian Estimation, and which does not imply a progression through time, we transfer the results from the Prediction to the Correction Phase:

$$\boldsymbol{x}_B = \boldsymbol{x}_M(t+1) \tag{8.3}$$

$$\boldsymbol{P}_B = \boldsymbol{P}_M(\boldsymbol{t}+\boldsymbol{1}) \tag{8.4}$$

This is just a "change of variable," which does not imply any calculation, and it is inserted here to clarify how the "outputs" of the Prediction Phase are used as "inputs" to the Correction Phase, OF THE CURRENT ITERATION (in spite of the fact that they appear with the "t + 1" time argument). In addition, we now use the subindex "B" (BEFORE) to emphasize that now $\mathbf{x}_B$ and $\mathbf{P}_B$ will be used in a process that does not imply a time progression from t to t + 1, but instead a refinement of the estimate that will transform these "prior" estimates (before) to enhanced "posterior" estimates (after) by means of Bayesian Estimation.

The Correction Phase (also called "Update Phase" in some texts) will use the $\mathbf{x}_B$ and $\mathbf{P}_B$ as the "prior estimates" of the state variables and their uncertainty, for the process of Bayesian Estimation. In this process the "added data" or "likelihood" will be provided by the current readings of the measurements $\mathbf{z}(t)$, which we will just describe as $\mathbf{z}$, as we saw in Chapter 6. Operationally, the "posterior" estimates are obtained by the following computations (re-written from Chapter 6):

"CORRECTION EQUATIONS" (or "UPDATE EQUATIONS"):

$$K_G = P_B H^T \left(H P_B H^T + R \right)^{-1} \tag{8.5}$$

$$x_A = x_B + K_G \left(z - H x_B \right) \tag{8.6}$$

$$P_A = P_B - K_G H P_B \tag{8.7}$$

In practical terms, we must first calculate the parameter (matrix) $\mathbf{K}_G$, so that it can be substituted in the next two equations. We notice that these three equations require the substitution of $\mathbf{H}$, the matrix that relates measurements to state variables as $\mathbf{z}(t) = \mathbf{H}(t)\mathbf{x}(t)$. If this relationship varies with time, we would need to provide the correct value of $\mathbf{H}(t)$ *for this point in time, specifically*. However, although the Kalman Filter framework allows for those cases, the relationship between measurements and state variables is often constant. In those cases, the same (constant) $\mathbf{H}$ matrix will be used in all iterations. Similarly, while the correlation matrix representing the uncertainty of the measurements, $\mathbf{R}(t)$, could theoretically change through time, in many cases the uncertainty factors affecting the measurements do not change through time. In those cases, the same (constant) matrix $\mathbf{R}$ will be used for all iterations.

In contrast, it is clear that each iteration of the Kalman Filter in its Correction Phase will require the involvement of the current values of the measurements, $\mathbf{z}(t)$, which we are simply representing in the Equation 8.6 as $\mathbf{z}$.

8.2 A BLOCK DIAGRAM AND A MATLAB® FUNCTION FOR IMPLEMENTATION OF ONE KALMAN FILTER ITERATION

The flow of information and calculations that take place during A SINGLE KALMAN FILTER ITERATION are displayed in Figure 8.1. The biggest, shaded rectangle represents the processor implementing the algorithm. The two smaller rectangles inside represent the two phases: Prediction and Correction. The vertical lines coming into the biggest rectangle represent the instantaneous data that may be required for the computation of this iteration of the algorithm. The lines for **F**, **G**, **Q**, **H** and **R** are dashed and lead to internal boxes with these variable names in them, because, if the systems are time invariant and the statistics of the variables involved are constant, these parameters will not need to be read in every iteration (they will be fixed values that only need to be initialized once).

The vertical line labeled "**z**" is solid because a newly acquired set of measurements must be read in every iteration, as they are necessary to implement the Correction Phase of the algorithm. The line for **u** is dashed

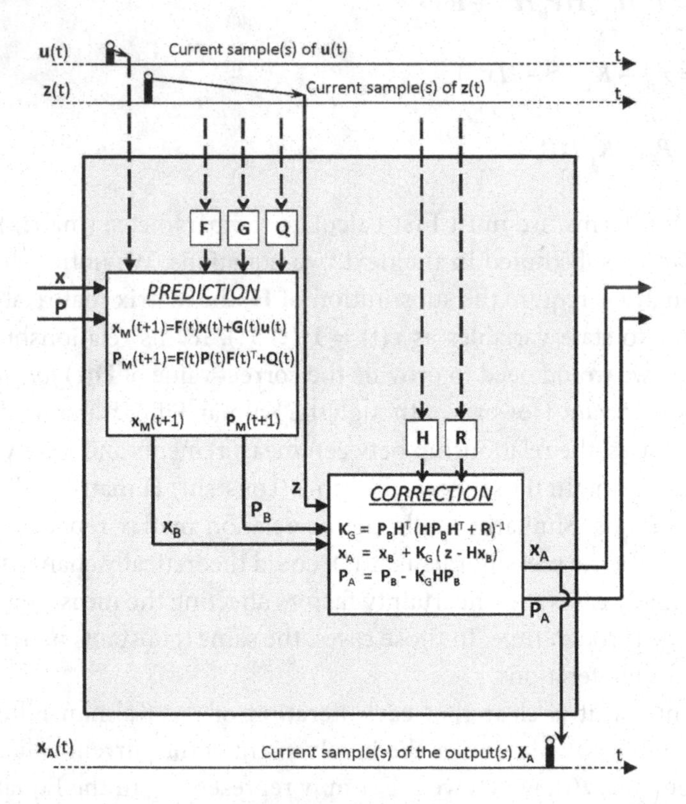

FIGURE 8.1 Block diagram illustrating the operations to be executed in one iteration of the Kalman Filter algorithm.

because, as we have explained before, some situations will not have a "control input" applied to the system.

The vertical line at the bottom of the biggest rectangle represents the fact that, in each iteration, the process will generate a final, enriched estimate of the state of the system, $\mathbf{x}_A$. (An estimate of $\mathbf{P}_A$ is also available at the end of every iteration.)

The horizontal lines coming into the process from the left represent the previous estimates of **x** and **P**, *obtained in the previous iteration of the algorithm (or provided as initial values).* The horizontal lines leaving the biggest rectangle on the right represent the posterior values x_A and P_A which will be passed on to the next iteration of the Kalman Filter.

MATLAB® function "onedkf.m" implements this single one iteration of the Discrete Kalman Filter. The comments included identify which lines implement which of the seven equations presented at the beginning of this chapter (Eq. 8.1 to Eq. 8.7).

```
%%% MATLAB CODE 08.01 ++++++++++++++++++++++++++++++++
% % % % % % % % % % % % % % % % % % % % %
% Function onedkf—Implements A SINGLE ITERATION OF
% THE DISCRETE-TIME KALMAN FILTER ALGORITHM
%
% SYNTAX: [PA, xA, KG] = onedkf(F,G,Q,H,R,P,x,u,z);
% Uses(receives) matrices F, G for the model equations
% Uses(receives) the process noise covariance matrix, Q.
% Uses(receives) matrix H for the measurement equation.
% Uses(receives) the measurement noise cov. matrix, R.
% The following are expected to change in every
% iteration:
% Receives the state vector, x, and its cov. matrix, P,
% from the previous iteration of the algorithm.
% Receives the current vector of inputs, u.
% Receives the current vector of measurements, z.
% Performs the prediction and Correction Phases.
% Returns the POSTERIOR estimation of state vector, xA
% and its covariance matrix, PA.
% It also returns the calculated KG matrix.
%
function [PA, xA, KG] = onedkf(F,G,Q,H,R,P,x,u,z);

%%% PREDICTION PHASE, using the MODEL:
% Equation (8.1)- The MODEL predicts the new x:
FT = transpose(F);
```

```
xM = F * x + G * u ;

% Equation (8.2)- The MODEL predicts the new P:
PM = F * P * FT + Q ;

%% Chge. of variables, to clearly separate the 2
phases:
xB = xM ; % Equation 8.3
PB = PM ; % Equation 8.4

%%% CORRECTION (UPDATE) PHASE
% -Finding POSTERIOR (A = After) parameters,
% from PRIOR (B = before), through Bayesian
% estimation:
HT = transpose(H);
% First calculate the Kalman Gain, KG, for this iteration
% (Equation 8.5):
KG = PB * HT * ( inv( H * PB * HT + R) ) ;

% Equation (8.6)-Calculate POSTERIOR ESTIMATE of the
% state vector
xA = xB + KG * (z-H * xB) ;

% Equation (8.7)-Calc. POSTERIOR ESTIMATE of state
% vector's covar. mtx.
PA = PB-KG * H * PB;

end
% % % % % % % % % % % % % % % % % % % %
%%% MATLAB CODE 08.01 ++++++++++++++++++++++++++++++++
```

8.3 RECURSIVE EXECUTION OF THE KALMAN FILTER ALGORITHM

Of course, to continue to provide enriched (posterior) estimates of $\mathbf{x}_A$ and $\mathbf{P}_A$ as time goes on, this function needs to be executed recursively. The resulting $\mathbf{x}_A$ and $\mathbf{P}_A$ from one iteration are fed forward as the $\mathbf{x}(t)$ and $\mathbf{P}(t)$ for the Prediction Phase of the next iteration. For the execution of the very first iteration there will not be "previous" $\mathbf{x}_A$ and $\mathbf{P}_A$ available. For that first iteration initial values must be provided which are defined from the situation being considered.

Figure 8.2 represents the consecutive implementation of Kalman Filter iterations described previously. In this figure each of the shaded rectangles

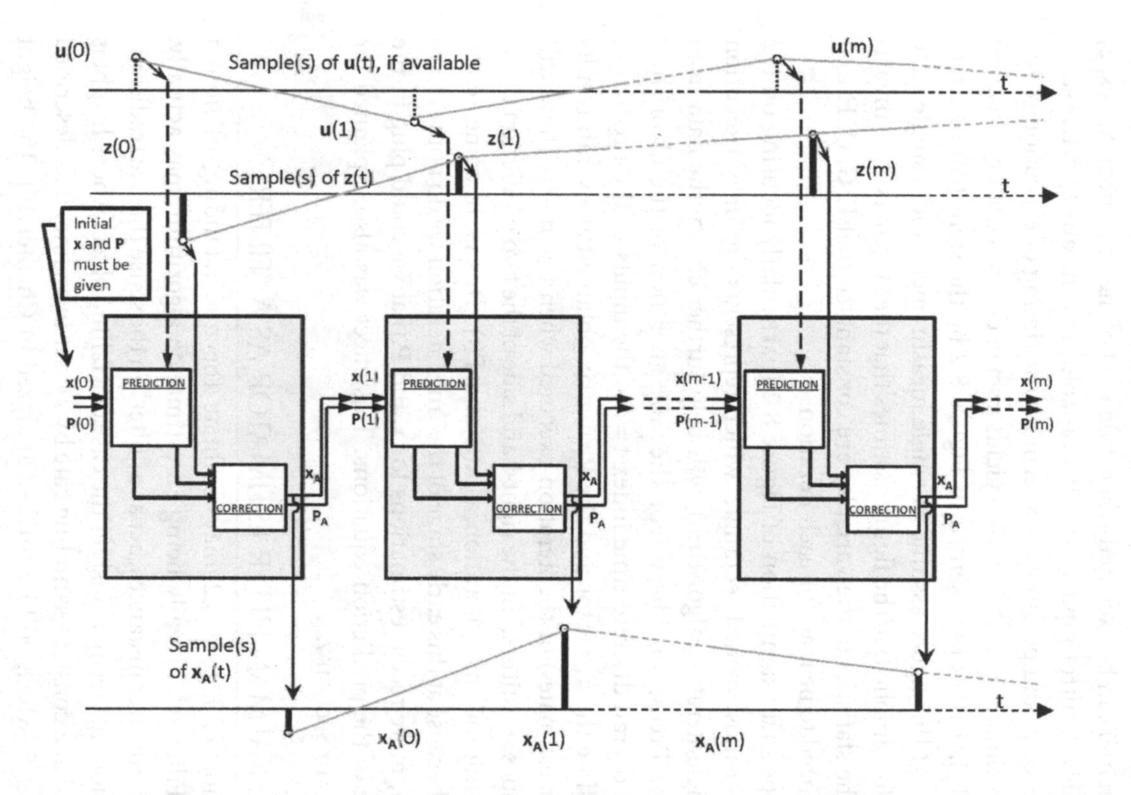

FIGURE 8.2 Visualization of the recursive operation of the Kalman Filter algorithm. Each shaded rectangle represents a single iteration of the algorithm, which is internally described as in Figure 8.1. During each iteration the algorithm reads in new data, **z** (and possibly new control inputs, **u**), as shown in the top area of the figure. In each iteration, the algorithm generates a posterior state vector, $\mathbf{X}_A$, and its associated covariance matrix, $\mathbf{P}_A$, as shown at the bottom of the figure.

contains all the elements of the single shaded rectangle in Figure 8.1. Prior to the very first execution of the algorithm, initial values $\mathbf{x}(0)$ and $\mathbf{P}(0)$ must be provided for the Prediction Phase. Thereafter, the $\mathbf{x}_A$ and $\mathbf{P}_A$ from the previous iteration are used in the Prediction Phase. Each iteration ("iteration m") can be executed when a new set of measurement values, **z**, and possibly "control inputs," **u**, are available. That means that the sampling rate for **z** (and possibly **u**) determine the timing of execution of the algorithm. Similarly, each iteration yields a set of estimated state variables, $\mathbf{x}_A$, (and $\mathbf{P}_A$). That is represented in Figure 8.2 by the stem plot of an $\mathbf{x}_A$ time series. (This figure assumes a single measurement and a single state variable, for simplicity.) The figure assumes that the system is time invariant and the statistics of the variables are constant, so that **F**, **G**, **Q**, **H** and **R** do not need to be read for each iteration.

We hope that examination of Figure 8.2 (particularly focusing on the third, right-most shaded rectangle, which represents the "m^{th}" execution of the single-iteration algorithm), will help further clarify the data used during the Prediction Phase. For the "m^{th}" iteration of the algorithm, when the overall discrete-time index $t = m$, the inputs to the Prediction Phase will be the $\mathbf{x}_A$ vector and the $\mathbf{P}_A$ matrix obtained as overall results (posterior estimates) of the iteration executed when $t = m - 1$. These will be the values substituted in the right-hand side of the two prediction equations during the "m^{th}" iteration as $\mathbf{x}(t)$ and $\mathbf{P}(t)$. This is, in some sense, obvious, because at this early stage of the "m^{th}" iteration we do not yet have any other, more recent estimations for **x** and **P** that we could plug in the right-side of the prediction equations. This effect was also explained for the reader in Section 7.3.

8.4 THE KALMAN FILTER ESTIMATOR AS A "FILTER"

Observation of Figure 8.2 may facilitate the understanding of how a Kalman Filter, while really being an estimation algorithm, may actually, under certain circumstances, act as a "filter" in the sense that we traditionally envision that term. Consider the case of a falling object whose height is monitored at a constant sampling rate by a laser instrument, as described in Chapter 5 (which will be further analyzed in Chapter 10). The height readings from the instrument will be our z measurement, which is, nonetheless subject to measurement noise. On the other hand, the assumption of a constant and known gravity and the potential influence of other factors allow us to set up an analytical model for the height through time (as we did in Section 5.3), which will be part of the state variables for this

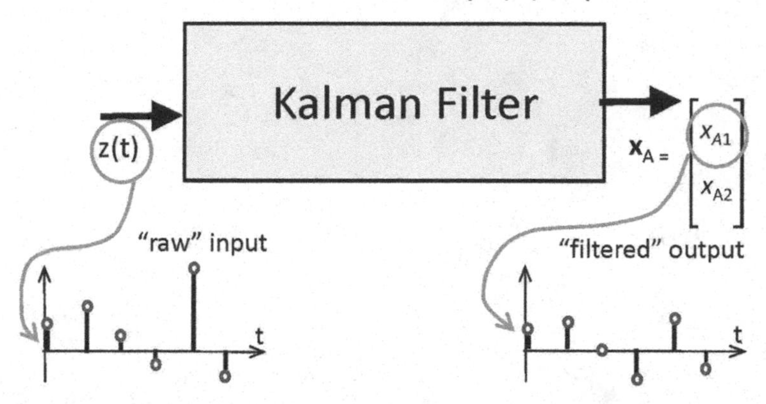

FIGURE 8.3 Illustration of a specific univariate case in which the Kalman Filter estimator may be seen as a "traditional filter," receiving in each iteration a scalar range measurement, z, and yielding a "filtered" output sample (the first element of the $\mathbf{x}_A$ vector for that same iteration), representing an "enhanced" estimation of the height of the falling object.

system, $\mathbf{x}(t)$. But the uncertainty of the model will likely cause the height estimations obtained from the model alone to depart from reality.

If the Kalman Filter is implemented recursively, as shown in Figure 8.2, every time a new sample of the measured height is read in z, the algorithm will generate a posterior estimate $\mathbf{x}_A$. Each one of those posterior state variable vectors will contain a corresponding "filtered," i.e., enhanced, estimate of the height (called "y_k" in Section 5.3) for that same time, which will have the added benefit of containing information provided by the model. From this perspective, the Kalman Filter algorithm will be implementing an "online filtering" functionality that matches our traditional concept of "filtering," where a raw signal is fed into a "filter" and an enhanced ("filtered") version of that signal is obtained at the output of the filter, as diagrammed in Figure 8.3.

CHAPTER 9

Univariate Example of Kalman Filter in MATLAB®

In this chapter we will apply the MATLAB® function that was developed in Chapter 8 to implement each iteration of the Kalman Filter to the simple, univariate scenario we first described in Chapter 4. This scenario deals with the estimation of an internal, inaccessible voltage for which we can only observe the speed of rotation of a motor supplied by the voltage we are trying to estimate.

9.1 IDENTIFICATION OF THE KALMAN FILTER VARIABLES AND PARAMETERS

Recapping the scenario we presented in Chapter 4, we have a module that is connected to a car battery with a (nominal) voltage VBATT = 12.0 V. Internally, VBATT drives a single loop circuit comprising two 10 kΩ resistors, R1 and R2, and a Zener diode (with Vz = 5.1 V). The effective voltage that will be present between R1 and R2 is the only state variable, $x_1 = V_x$, and the only measurement available is the rotating speed, r, of a DC motor driven by V_x, which is specified so that $r = 5\ V_x$, in RPMs. The electric circuit for this scenario was shown in Figure 4.1 (imprinted on the side of the hypothetical device).

According to the circuit analysis presented in Chapter 4, we recall Equation 4.3:

$$V_x = Vz + iR2 = Vz + \frac{(VBATT - Vz)R2}{R1 + R2} \tag{9.1}$$

Using the nominal values already presented:

$$V_x = (5.1V) + \frac{(12V - 5.1V)(10{,}000)}{10{,}000 + 10{,}000} \tag{9.2}$$

$$V_x = 8.55V \tag{9.3}$$

In this system we don't really expect a dynamic evolution (of the system) through time, and therefore the model predicts that this value will remain unaltered. So, the model equation is simply:

$$\boldsymbol{x}(t+1) = \boldsymbol{x}(t) \tag{9.4}$$

Since there will be likely departures from the nominal values of the components and due to the actual loading of the DC motor, we will consider that:

1. The external model noise will be set to have a variance $Q = \sigma^2 = 0.001$.
2. The uncertainty of the initial state estimate will be set to $P(0) = 0.3$.
3. We will propose $x_1(0) = 8.55$ V, for the initial value of the voltage, based on the result obtained from the model under ideal circumstances (Equation 9.3).

The only variable that is being measured is the speed of rotation of the motor axis: $r = 5\ V_x$, in RPMs. Therefore $\mathbf{H} = h = 5$, and we will be using a sampling interval of 1 minute (and we will assume we get rotational speed readings every minute). We will assume an uncertainty for our speed of rotation readings of $R = \sigma_r^2 = 5.0$.

9.2 STRUCTURE OF OUR MATLAB® SIMULATIONS

The function onedkf developed in Chapter 8 implements only one iteration of the Kalman Filter algorithm, requiring the values of $\mathbf{x}(t)$ and $\mathbf{P}(t)$ (which are the $\mathbf{x_A}$ and $\mathbf{P_A}$, from the previous iteration) and yielding the posterior estimation results $\mathbf{x_A}$ and $\mathbf{P_A}$ for the current iteration.

It is clear then that for the processing of an actual time series in a given scenario we will need to execute onedkf recursively. We will achieve this by calling onedkf within a function that contains a timing loop, where a control

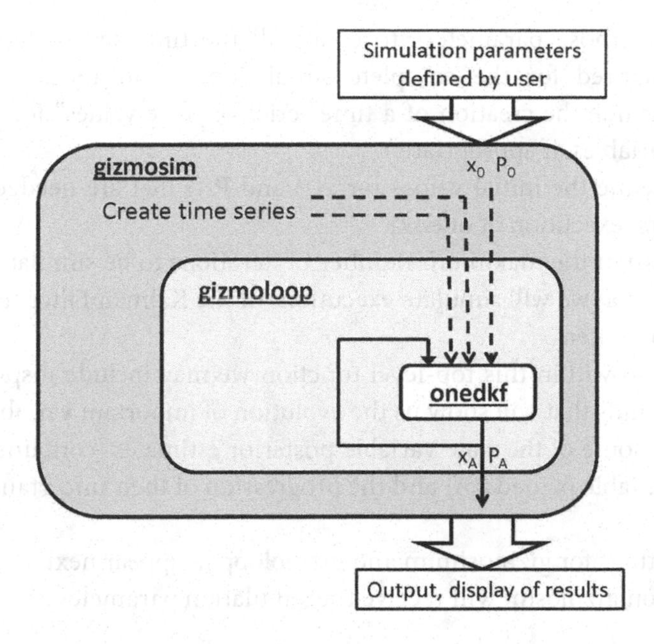

FIGURE 9.1 General simulation structure for the estimation of the inaccessible voltage in the gizmo.

variable, "t," will act as the discrete-time index, progressing from t = 1 to a maximum number of iterations defined by us, "iter." For the case of estimating the inaccessible voltage in the gizmo first described in Chapter 4, we will create the timing loop function called "gizmoloop." (In general, we will try to name these timing loop functions "such-and-such-*loop*.")

Further, in every iteration of the loop, the Kalman Filter algorithm needs to draw new samples from the measurements (and other variables, such as control inputs, and parameters, if they are not constant). Therefore, we will embed the loop function within a top-level simulation function, in which we will create, in advance, all the time series that will be required, in their entirety, according to our modeling of the situation at hand and following the modeling parameters we want to simulate. For the estimation of the inaccessible voltage inside the gizmo, we will call this top-level function "gizmosim." (In general, we will try to name these top-level simulation functions "such-and-such-*sim*.") The relationship between gizmoloop and gizmosim is shown in Figure 9.1. This top-level function will enable us to:

1. Define the parameters of the model we want to use, such as the levels of uncertainty that we want to simulate for the external noise (**Q**) and the measurements (**R**).

2. Use those parameters to create all the time series that will be required for the complete simulation, in advance. (This may include the creation of a time series of "true values" for the state variables, if appropriate.)
3. Define the initial values for **x**(t) and **P**(t) that are needed for the first execution of onedkf.
4. Assign the maximum number of iterations to be simulated, "iter," so that we will simulate executions of the Kalman Filter from t = 1 to t = iter.
5. Also within this top-level function we may include display commands that will show us the evolution of important variables, such as some of the state variable posterior estimates (contained in the variable named `xA`) and the progression of their uncertainties.

The listings for gizmosim.m and gizmoloop.m appear next.
Function gizmosim will receive the simulation parameters:

- `xtru`: (Scalar) value that will be used as the mean of the "true" values of the inaccessible voltage for the simulation.
- `x0`, `P0`: Initial values that will be used in the first execution of onedkf. Because there is only one state variable in this scenario, `P0` will be a scalar representing the initial model uncertainty for that variable, expressed as its variance.
- `Q`, `R`: Covariance matrices (in this case scalar variances) which represent the uncertainty of the model attributed to external inputs (noise) and the uncertainty in the rotational speed measurements.
- `iter`: number of total iterations we want to simulate (from t = 1 to t = iter).

9.3 CREATION OF SIMULATED SIGNALS: CORRESPONDENCE OF PARAMETERS AND SIGNAL CHARACTERISTICS

With these parameters gizmosim will create the necessary time series and setup the simulation, so that the timing loop, gizmoloop, can then be executed.

This first, univariate example is a great opportunity for us to reflect on the role that the Kalman Filter parameters play in both, our modeling of the phenomenon and the execution of the Kalman Filter algorithm. Since this is a simulation, it will be advantageous to define a time series that we can consider to be the "true" voltage V_x throughout the analysis time. Having this time series available will allow us to perceive how close or how far our Kalman Filter output comes to the "true" state variable. So, what will be the voltage in the node that connects those two resistors? Our model is a static, time-invariant electrical system. Therefore, we would not expect the model to predict changes in any of the voltages around the loop. However, we realized that we must acknowledge the presence of external influences, which will likely exist in reality. While we do not contemplate any external "control inputs" (i.e., u(t)) applied to this system, we must acknowledge that there will be, nonetheless, external noise acting on the voltage of that node, due to phenomena that we do not control, but that are likely present. That is why we are adopting a non-zero value of **Q** in this example (and use that value in the Kalman Filter computations). Because this is a univariate case, **Q** will be a 1-by-1 covariance matrix, that is, it will be a scalar with a value defined by the variance of the external noise. (Another way to think about this is to consider that we are dealing with a u(t) that has NO MEAN VALUE and it is only involuntary, external noise, with a variance whose value will be used as (a scalar) **Q**.) Therefore, our simulated "true" voltage between the resistors will be a certain mean value added with a time series of Gaussian random values with zero mean and scaled to have a standard deviation that is the square root of the **Q** parameter for our situation. The resulting time series (mean value plus random noise time series), is created in the "`xtrue`" vector in the MATLAB® code to follow.

To perform the first iteration of the Kalman Filter we will need to provide initial values for **x**(t) and its covariance matrix, **P**(t). In this univariate case we only need a scalar "initial estimate" of the voltage between the two resistors. It seems logical to propose x(0) = 8.55 V, since that is the value that our analysis of the electric circuit, under ideal circumstances, would indicate. However, in a real scenario circumstances will not be ideal. It is likely that there will be loading effects that are not represented in our basic analysis of the circuit. Because of that, we will choose to assume a mean value for the "true" voltage time series that is lower than 8.55 V. For example, we could propose for the simulation that the mean value of the "true" voltages is 8.0 V. For the initialization of the Kalman Filter we also need to propose an initial P(0), which in this case is just a scalar variance representing the level of uncertainty we assign to the initial state estimate we are proposing, namely, x(0) = 8.55 V.

In addition to the time series and initial values mentioned earlier, the simulation will also require the creation of a time series of measurements, z(t) = r(t), which stands for the series of rotational speeds that we would measure (every minute) from the rotating shaft (in RPMs). According to the relationship between the inaccessible voltage and the rotational speed, we would expect that the time series of measurements could ideally be just "xtrue" times 5. However, we are acknowledging measurement error with a standard deviation that is the square root of the only value in R for this scenario. Therefore, we will create the time series z(t) by magnifying the "xtrue" time series by a factor of 5, and then superimposing on this partial result Gaussian noise samples with 0 mean and a standard deviation that is the square root of R.

Summarizing, gizmosim creates a vector xtrue of length iter, as described earlier. Then it creates a vector of measured rotation speed values, zvect, by multiplying the true values by 5 (since h = 5 in this case) and superimposing normally distributed random (measurement) noise with a standard deviation adjusted to be the square root of the uncertainty of the only measurement variable.

```
%%% MATLAB CODE 09.01 ++++++++++++++++++++++++++++++++
% gizmosim.m—top-level function for simulation of
univariate Kalman Filter
% which estimates an inaccessible voltage from rotation
% speed in the hypothetical gizmo described in Chapter 4
%
% SYNTAX: [XAVECT, PAVECT, KGVECT] = gizmosim(xtru, x0,
% P0, Q, R, iter);

function [XAVECT,PAVECT,KGVECT] = gizmosim(xtru,x0,P0,
Q,R,iter);

rng(12345,'v5normal'); % Resets the Random Number
% Generator to ver5. Normal, seed = 12345;


% No u, but there is external noise represented by
% Q = variance of noise
% Creation of vector of TRUE voltage values:
xtrue_nonoise = ones(iter,1) * xtru;
extnoise = randn(iter,1) * ( sqrt(Q) );
xtrue = xtrue_nonoise + extnoise;

% Creation of vector of MEASURED values, taking
% into account that the measuring system ADDS noise to
```

```
% each TRUE VALUE measured:
measerror = randn(iter,1) * (sqrt(R));
zvect = (xtrue * 5) + measerror;

[XAVECT,PAVECT,KGVECT]=gizmoloop(xtrue,zvect,x0,P0,Q,
R,iter);
zvecdiv5 = zvect ./5 ;
gray6 = [0.6, 0.6, 0.6];

figure;plot(xtrue,'Color',gray6,'LineWidth',1.5);hold on;
plot(XAVECT,'k','LineWidth',1.5 );
plot(zvecdiv5,'k-.','LineWidth',1.5);
hold off; title('xtrue, xA and z/5');axis([0,100,0,9]);
xlabel('Kalman Filter Iterations');
legend('xtrue','xA','zvecdiv5','Location','southeast');
grid off;

figure; plot(PAVECT,'k','LineWidth',1.5);title('PA');grid;
xlabel('Kalman Filter Iterations');
figure; plot(KGVECT,'k','LineWidth',1.5);
title('KG values');grid;
xlabel('Kalman Filter Iterations');

end
%%% MATLAB CODE 09.01 ++++++++++++++++++++++++++++++++
```

9.4 THE TIMING LOOP

Function gizmoloop receives the initial values in variables x0, P0, and loads them in the x, P variables that will be passed to onedkf. It also passes along the parameters Q and R, which will remain constant. It receives both time series xtrue and zvect. The vector zvect will be accessed by onedkf in every iteration. Function gizmoloop also receives the value of iter, to determine the maximum value for the loop control variable t (maximum number of iterations to be simulated).

In addition, function gizmoloop establishes the parameters of the scenario (model and measurement equations), assigning values to **F** and **H**, which will remain constant throughout the iterations. In this case **G** is assigned a value of 0, as no intentional control inputs are contemplated. This function also prepares storage to save the state variable estimate in the variable named xA and its corresponding uncertainty, in the variable PA (which is a scalar in this case, representing the variance of xA), as well as the Kalman Gain (a scalar, in this case), from every iteration simulated. This will allow us to display the evolution of these three values through time when all the iterations have been simulated.

```
%%% MATLAB CODE 09.02 ++++++++++++++++++++++++++++++++
% gizmoloop.m—timing loop for simulation of univariate
% Kalman Filter for the estimation of an inaccessible
% voltage from rotation speed in the hypothetical
% gizmo described in Chapter 4
%
% SYNTAX:[XAVECT, PAVECT,KGVECT]
% =gizmoloop(xtrue,zvect,x0,P0,Q,R,iter);

function [XAVECT, PAVECT, KGVECT]=gizmoloop(xtrue,zvec
t,x0,P0,Q,R,iter);

F = 1;
G = 0;
H = 5;

x = x0;
P = P0;

% Input time series
u = zeros(iter,1);
z = zvect;

% Set up vectors to store outputs (all iterations)
PAVECT = zeros(iter,1);
XAVECT = zeros(iter,1);
KGVECT = zeros(iter,1);

for t = 1:iter %%% ----- START OF TIMING LOOP

  [PA, xA, KG] = onedkf(F,G,Q,H,R,P,x,u(t),z(t));
  %[PA, xA, KG] = onedkfkg0(F,G,Q,H,R,P,x,u(t),z(t));

  PAVECT(t) = PA;
  XAVECT(t) = xA;
  KGVECT(t) = KG;
```

```
  % pass results as inputs to the NEXT iteration:
  P = PA;
  x = xA;

end       %%% ----- END OF TIMING LOOP

end
%%% MATLAB CODE 09.02 ++++++++++++++++++++++++++++++++
```

9.5 EXECUTING THE SIMULATION AND INTERPRETATION OF THE RESULTS

Figures 9.2, 9.3 and 9.4 show the results we obtain by simulating with the following parameters, which are assigned typing these statements in MATLAB®'s command window:

```
xtru = 8 % (Simulating that the true mean value of
% the voltage is actually only 8.0 V, due to loading)
x0 = 8.55 % (Using the value obtained from the model
% with nominal component values, Equation 9.3)
P0 = 0.3 % (Assigning an initial variance of 0.3
% volts-squared for the results from the model)
Q = 0.001 % (Assigning a variance, of 0.001
% volts-squared for the external noise in the model)
R = 5 % (Assigning a variance of 5 to the speed
% measurements)
iter = 100 % (Will simulate 100 iterations,
% from t=1 to t=100).
```

Then, we invoke gizmosim in this way:

```
[xAVECT,PAVECT,KGVECT]=gizmosim(xtru,x0,P0,Q,R,iter);
```

Figure 9.2 compares the estimates found by the Kalman Filter (solid black trace), with the values of r measured at each iteration, divided by 5 (dashed trace). In this univariate case, r = hx = (5)(x). Therefore, it is possible to calculate the values of the state variable x that would be derived exclusively from the measured rotational speeds as x = (r)(1/5). We see that, if we were to rely exclusively on the information from the measurements, our estimate of the inaccessible voltage would not be impacted by the wrong initial value assigned x0 = 8.55 V, as the dashed trace

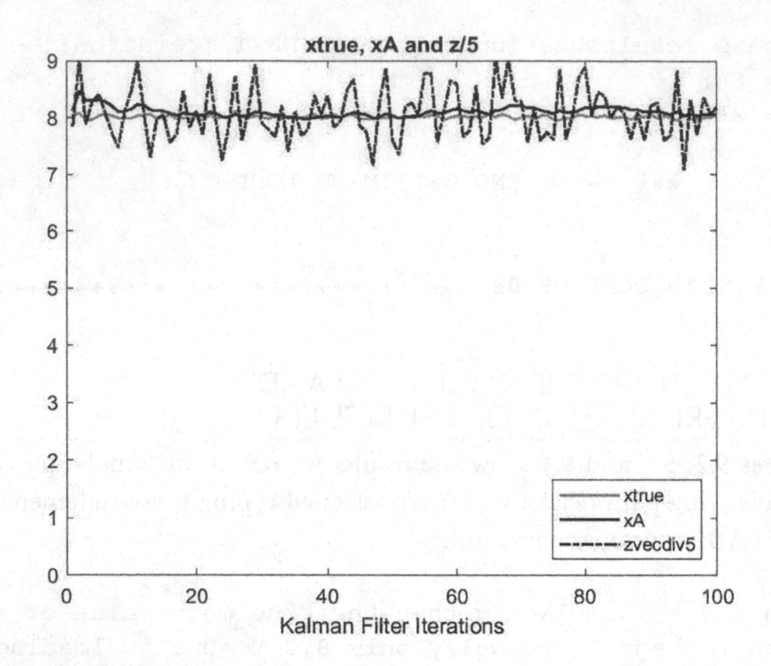

FIGURE 9.2 Evolution of the "true" V_x voltage, z/5 and the posterior estimate of the state variable, x_A, through 100 iterations of the Kalman Filter algorithm.

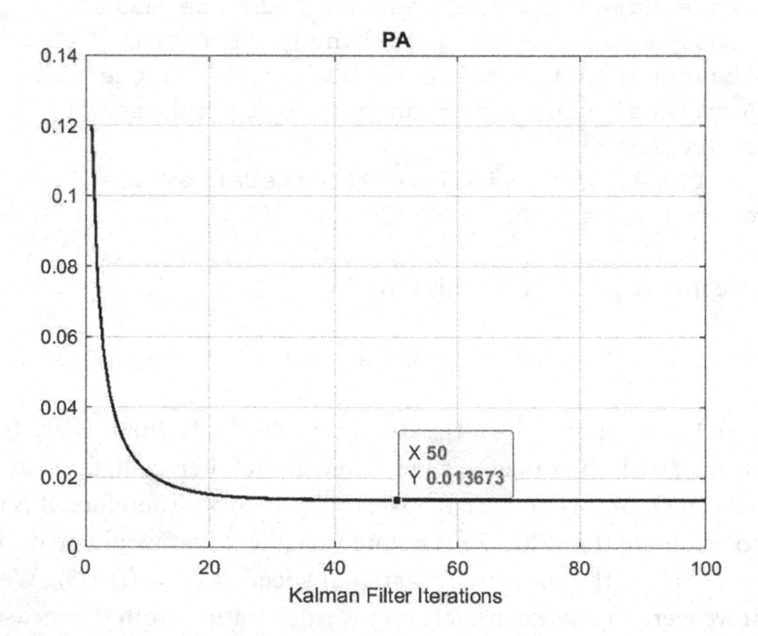

FIGURE 9.3 Evolution of the (scalar) P_A, (i.e., the variance of x_A), through 100 iterations of the Kalman Filter algorithm.

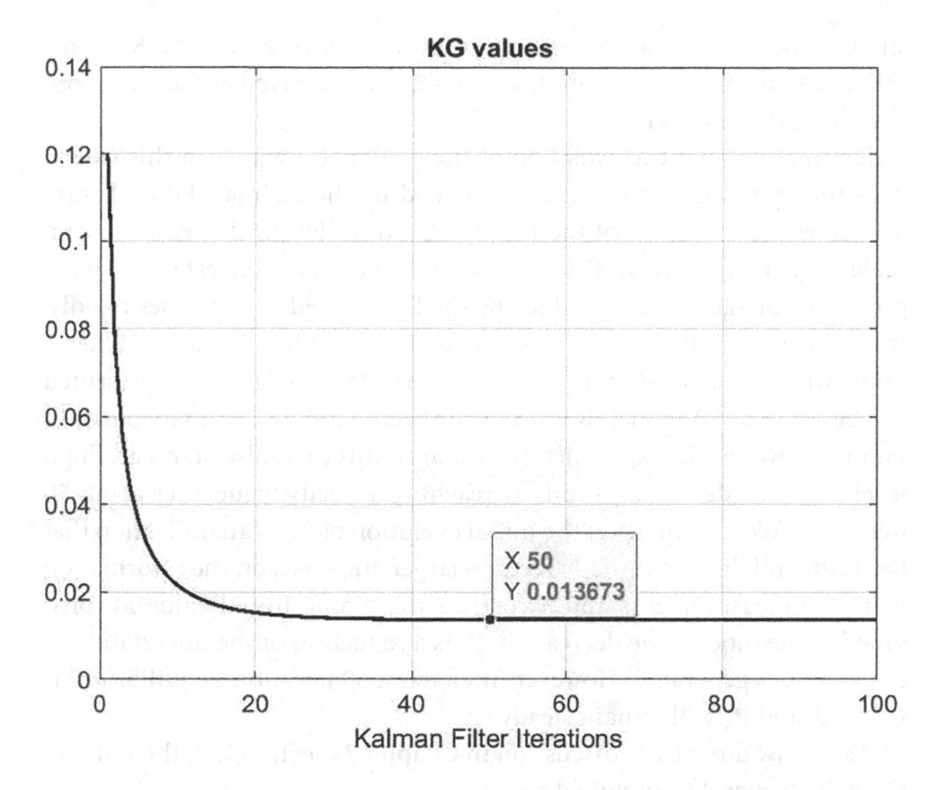

FIGURE 9.4 Evolution of the (scalar) Kalman Gain, K_G, through 100 iterations of the Kalman Filter algorithm.

fluctuates about the correct real value of 8.0 V. However, this empirical estimate is "very noisy," with the transient fluctuations of the dashed trace reaching as low as close to 7.0 V and as high as almost 9.0 V (due to the high value of R = 5). The estimates provided by the Kalman Filter in xA (solid black trace), quickly approach the true mean value simulated (8.0 V), and, in addition show smaller fluctuations. That is, a "noise level reduction" has been achieved from the estimate exclusively based on measurements, to the xA results provided by the Kalman Filter. (It should be noted that *it is not always possible* to directly estimate x values from the measurements, if, for example, matrix H does not have an inverse, in the multivariate case.)

It is worth pointing out that we see an example of how, although it is really an estimator, the Kalman Filter is having the effect we commonly associate with a "filter." This is apparent in this case because of the much smaller "noise" we see in the voltage estimate yielded by the Kalman-Filter, `xA`, in comparison with the time sequence derived exclusively from the measurements as r / 5.

Figure 9.3 shows the evolution of the (scalar) P_A, which in this case is only the variance of the voltage estimated by the Kalman Filter. Figure 9.4 shows the evolution of the Kalman Gain, which in this case is just a scalar, k_G. We observe, in Figure 9.3, that the variance (uncertainty) of the posterior voltage estimate yielded by the Kalman Filter decreases rapidly, from the initial value `P0` we assigned, to a value of 0.01367 after 50 iterations, where it seems to stabilize. In this case, that variance is represented by the value of `PA`, which is a scalar for this example. Simultaneously, in Figure 9.4 we see that k_G, which is a scalar in this case, also starts at a high level and then decreases, similarly reaching a steady value after about 50 iterations. We can interpret the initial evolution of k_G as an indication that the Kalman Filtering initially favors a larger emphasis on the information from the measurements and it corrects the wrong initial value `x0` provided to the model. The decrease in P_A is a reduction of the uncertainty of the x_A values generated. However, after about 50 iterations, equilibrium is reached, and P_A will remain steady.

As we saw during the discussion in Chapter 7 (Section 7.3), the Kalman Gain, $\mathbf{K}_G$, generally computed as:

$$\boldsymbol{K}_G = \boldsymbol{P}_B\boldsymbol{H}^T\left(\boldsymbol{H}\boldsymbol{P}_B\boldsymbol{H}^T + \boldsymbol{R}\right)^{-1} \tag{9.5}$$

will end up being a scalar for this case, since: $\mathbf{P}_B = \sigma_B^2$, $\mathbf{H} = h$, $\mathbf{R} = \sigma_z^2$; therefore, the scalar k_G, is defined as:

$$k_G = \frac{h\sigma_B^2}{h^2\sigma_B^2 + \sigma_z^2} \tag{9.6}$$

After the first iteration, the value of P(t) fed into a given iteration will be the same as the P_A result from the previous iteration (as we previously showed, in Figure 8.2 in Chapter 8). In Figures 9.3 and 9.4, we notice that, after 50 time steps, the values of P_A and k_G have adopted the following steady values: $P_A = 0.01367$ and $k_G = 0.01367$. In this state of equilibrium, each iteration starts by receiving the value of P_A and using it as P(t) for the model equation (Equation 6.34), reproduced here:

$$\mathbf{P}(t+1) = \mathbf{F}(t)\mathbf{P}(t)\mathbf{F}(t)^T + \mathbf{Q}(t) \tag{9.7}$$

for this particular univariate case, this implies, once the steady-state has been reached:

$$P_B = \sigma_B^2 = \mathbf{P}(t+1) = (1)(0.01367)(1)^T + 0.001 = 0.01467 \quad (9.8)$$

And, indeed:

$$k_G = \frac{h\sigma_B^2}{h^2\sigma_B^2 + \sigma_z^2} = \frac{(5)(0.01467)}{(25)(0.01467)+5} = 0.013667489 \approx 0.01367 \quad (9.9)$$

Then, with these values of k_G and P_B, the resulting P_A, *for this iteration*, will be (using Equation 6.39):

$$\boldsymbol{P_A = P_B - K_G H P_B} \quad (9.10)$$

$$P_A = 0.01467 - (0.01367)(5)(0.01467) = 0.013667305 \approx 0.01367 \quad (9.11)$$

which confirms that a steady state has been reached.

Overall, we can observe that the x_A estimates have benefited from the measurements, which have led them to adopt values close to the true mean value of 8.0 V. The x_A estimates have also benefited from the model, leading them to have much smaller departures from the mean true value (they are "less noisy") than those observed in the trace representing z/5 (noisier).

The reader can use gizmosim to simulate different variations of this scenario, with alternative values of x0, P0, Q, R, etc.

9.6 ISOLATING THE PERFORMANCE OF THE MODEL (BY NULLIFYING THE KALMAN GAIN)

Before we leave this scenario, we would like to show what would happen if we were to void the Correction Phase of the Kalman Filter process, ignoring the information provided by the measurements and relying exclusively on the estimation of the inaccessible voltage from the model. From observation of the last two equations that define the Kalman Filter, Equations 6.38 and 6.39, reproduced here:

$$\boldsymbol{x_A = x_B + K_G(z - Hx_B)} \quad (9.12)$$

$$\boldsymbol{P_A = P_B - K_G H P_B} \quad (9.13)$$

we notice that, if we overwrite the contents of matrix $\mathbf{K}_G$ (regardless of its dimensions) with all zeros, the second terms on the right-hand side of both equations will be cancelled and no effective correction will take place. That is, in that anomalous case $\mathbf{x}_A$ would simply be $\mathbf{x}_B$, and $\mathbf{P}_A$ would simply be $\mathbf{P}_B$, which means we would be forcing the Kalman Filter to just rely on the model, ignoring the information brought to bear by the measurements.

Just for the purpose of comparison, we have created an "anomalous" version of function onedkf in which we implement this bypassing of the Correction Phase, simply by adding this line:

```
"KG=zeros(size(KG));%->OVERWRITES KG WITH MATRIX OF
% ZEROS"
```

right after the line that actually calculates the correct KG matrix:

```
" KG = PB * HT * (inv( H * PB * HT + R) );"
```

We have called the MATLAB® function where we added that extra line "onedkfkg0," where the last three characters ("kg0") remind us that in this "anomalous" implementation of the Kalman Filter iteration we are forcing all the values in KG to 0 and, therefore, we are bypassing the correction of the estimate that uses information from the measurements. The listing of the "anomalous" onedkfkg0 function appears in "MATLAB® Code 09.03."

```
%%% MATLAB CODE 09.03 ++++++++++++++++++++++++++++++++
% % % % % % % % % % % % % % % % % % % % % % %
% Function onedkfkg0 —Implements A SINGLE ITERATION OF
% THE DISCRETE-TIME KALMAN FILTER ALGORITHM
% FORCING KALMAN GAIN TO ZERO (MATRIX)
% SO THAT NO CORRECTION TAKES PLACE
% OUTPUT (xA, PA) IS ONLY FROM MODEL
%
% SYNTAX: [PA, xA, KG] = onedkfkg0(F,G,Q,H,R,P,x,u,z);
% Uses(receives) matrices F, G for the model equations
% Uses(receives) the process noise covariance matrix, Q.
% Uses(receives) matrix H for the measurement equation.
% Uses(receives) the measurement noise cov. matrix, R.
% The following are expected to change in every
%iteration:
% Receives the state vector, x, and its cov. matrix, P,
% from the previous iteration of the algorithm.
% Also receives the current vector of inputs, u.
```

```
% Also receives the current vector of measurements, z.
% Performs the prediction and the correction phases.
% Returns the POSTERIOR estimation of state vector, xA
% and its covariance matrix. PA.
% It also returns the calculated KG matrix.
%
function [PA, xA, KG] = onedkfkg0(F,G,Q,H,R,P,x,u,z);

%% PREDICTION PHASE, using the MODEL:
% Equation (8.1)-The MODEL predicts the new x:
FT = transpose(F);

xM = F * x + G * u ;

% Equation (8.2)- The MODEL predicts the new P:

PM = F * P * FT + Q ;

%% Chge. of variables, to clearly separate the 2 phases:
xB = xM ; % Equation 8.3
PB = PM ; % Equation 8.4

%%% CORRECTION (UPDATE) PHASE—Finding POSTERIOR
% (A = After) parameters, from PRIOR (B=before),
% through Bayesian estimation:
HT = transpose(H);
% First calculate the Kalman Gain, KG, for this iteration
% (Equation 8.5):
KG = PB * HT * ( inv( H * PB * HT + R) ) ;
KG = zeros(size(KG)); % % --> OVERWRITES KG
% WITH MATRIX OF ZEROS
                        % % --> "BYPASSES" ANY
CORRECTION OF MODEL
% Equation (8.6)—Calculate POSTERIOR ESTIMATE
% of the state vector
xA = xB + KG * (z—H * xB) ;

% Equation (8.7)—Calc. POSTERIOR ESTIMATE
% of state vector's covar. mtx.
PA = PB—KG * H * PB;

end
%%% MATLAB CODE 09.03 ++++++++++++++++++++++++++++++++
```

If we, temporarily, invoke function onedkfkg0 instead of the Normal function onedkf inside the gizmoloop function, and we run a new simulation with exactly the same parameters as before, the time series of the x_A that will be returned will be the estimates that the model, *without the Correction Phase*, would yield.

Figures 9.5 and 9.6 show the graphic output of a simulation run performed with the altered gizmoloop function.

Figure 9.5 shows that, if we were to rely only on the model, the initial mistaken value proposed for the estimated voltage, x0 = 8.55, would be presented as the result of the algorithm and remain constant throughout the simulation. This shows clearly that if the measurements are not allowed to enrich the posterior estimate of voltage, the model alone will not be able to recover from the inaccuracy of our initial estimate.

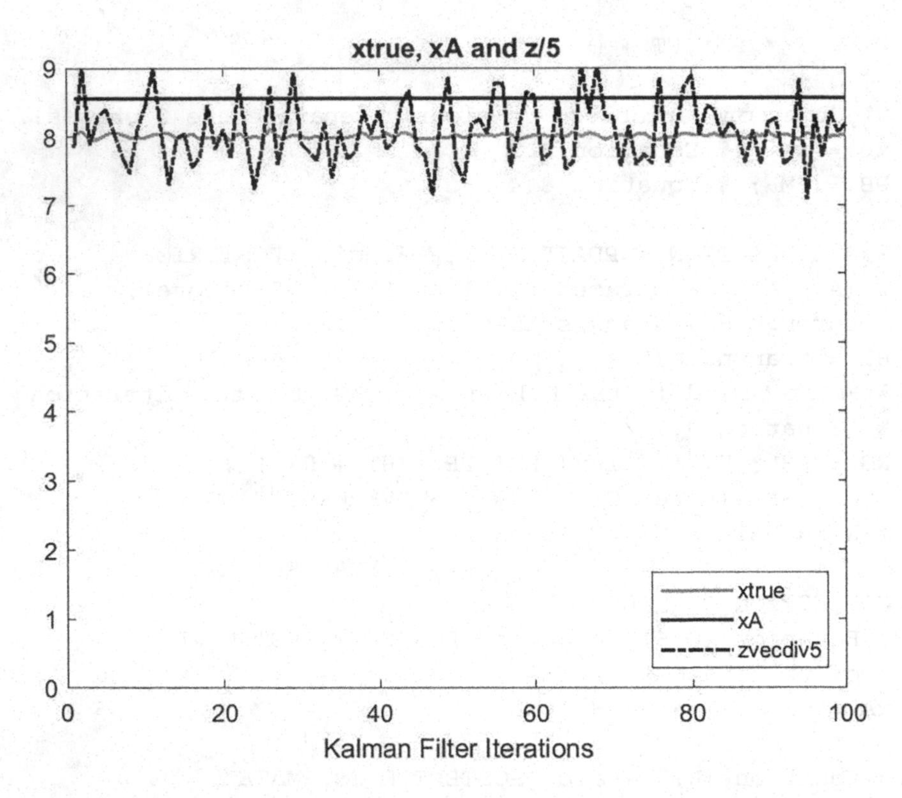

FIGURE 9.5 Evolution of the results from the Kalman Filter (solid black trace) when the Correction Phase has been bypassed. The output estimate stays constant at its initial value, unable to correct the inaccuracy in the initial estimate provided to the algorithm.

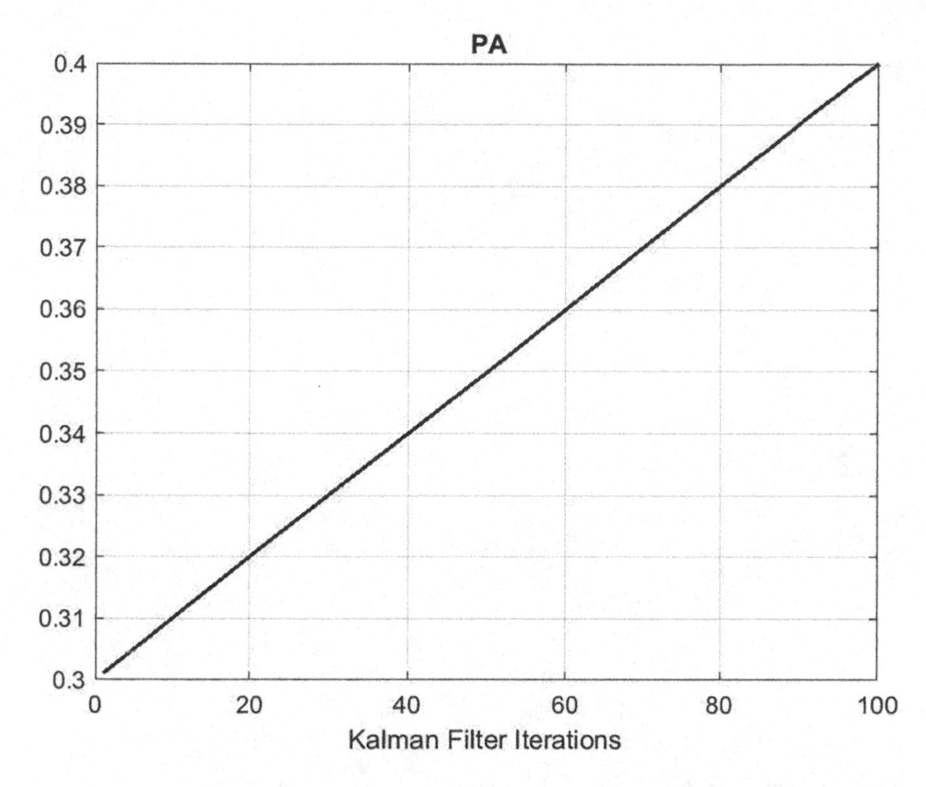

FIGURE 9.6 Evolution of the uncertainty of the state estimate (P_A) from the Kalman Filter when the Correction Phase has been bypassed. The variance of the estimate increases continuously, revealing that this anomalous implementation of the Kalman Filter is unable to reduce the uncertainty of the estimate it generates.

Figure 9.6, however, shows that, in this anomalous case, the Kalman Filter displays a steady increase of the uncertainty of the estimate that it is generating (P_A). This is not what we would expect and it could be interpreted as a "warning" that is signaling the inappropriate performance of the altered Kalman Filter algorithm.

CHAPTER 10

Multivariate Example of Kalman Filter in MATLAB®

In this chapter we will apply the MATLAB® function that was developed in Chapter 8 to address the scenario of the falling wad of paper presented in Chapter 5. In contrast with the case of the inaccessible voltage in our hypothetical gizmo, this case actually deals with a dynamic system and its evolution. Since there are two state variables, most of the parameters in the Kaman Filter algorithm are now matrices or vectors. It should be noted that the onedkf function developed in Chapter 8 can be used in this case too, since it was developed for the multivariate case and the univariate example studied in Chapter 9 was simply a scalar special case of the general process. We will visualize and reflect on the results from Kalman Filtering estimation of the height of the falling wad of paper.

10.1 OVERVIEW OF THE SCENARIO AND SETUP OF THE KALMAN FILTER

As detailed in Chapter 5, a discrete-time model for the evolution of the height of a falling wad of paper released from an initial height y_0, can be given by Equation 5.31 (or, generally Equation 6.1) with the assignments in Equations 5.28, 5.29 and 5.30, reproduced here:

$$\boldsymbol{x}(t+1) = \boldsymbol{F}(t)\boldsymbol{x}(t) + \mathbf{G}(t)\boldsymbol{u}(t) \tag{10.1}$$

$$\boldsymbol{x} = \begin{bmatrix} y_k \\ y'_k \end{bmatrix} \tag{10.2}$$

$$\boldsymbol{F} = \begin{bmatrix} 1 & \Delta T \\ 0 & 1 \end{bmatrix} \tag{10.3}$$

$$\boldsymbol{G} = \begin{bmatrix} -\left(\frac{1}{2}\right)(\Delta T)^2 \\ -\Delta T \end{bmatrix} \tag{10.4}$$

where ΔT is the sampling interval, i.e., the time between the recording of two consecutive samples of the signals involved, and g = 9.81 m/s^2 is assigned as **u**(t).

This yields the model presented in Equation 5.30, reproduced here:

$$\begin{bmatrix} y_{k+1} \\ y'_{k+1} \end{bmatrix} = \begin{bmatrix} 1 & \Delta T \\ 0 & 1 \end{bmatrix} \begin{bmatrix} y_k \\ y'_k \end{bmatrix} + \begin{bmatrix} -\left(\frac{1}{2}\right)(\Delta T)^2 \\ -\Delta T \end{bmatrix} g \tag{10.5}$$

In all of these equations, y is the instantaneous height of the wad of paper and y′ is its time derivative (i.e., its speed). The parameter g represents Earth's gravity. The model in Equation 10.5, however, does not represent the opposition that friction between the wad of paper and air will present to the falling motion. As we proposed in Chapter 5, we will model that variable opposition substituting the constant value g with a time-varying "diminished downwards acceleration." We will call this time-varying acceleration "actualg(t)." It will have a mean value below 9.81 m/s^2, given by g – g_{back}, where "g_{back}" is a positive parameter that we will set. In addition, actualg(t) will fluctuate above and below its mean value because we will add to the mean value a series of random values from a Gaussian distribution with 0 mean and a variance $\sigma^2_{actualg}$. This models the fact that the wad of paper will rotate as it falls and therefore cause variations in the opposing friction force.

After the substitution just described, the model equation would look like this:

$$\begin{bmatrix} y_{k+1} \\ y'_{k+1} \end{bmatrix} = \begin{bmatrix} 1 & \Delta T \\ 0 & 1 \end{bmatrix} \begin{bmatrix} y_k \\ y'_k \end{bmatrix} + \begin{bmatrix} -\left(\frac{1}{2}\right)(\Delta T)^2 \\ -\Delta T \end{bmatrix} actualg_k \tag{10.6}$$

Further, to put the equation in a format that is very close to the format in Kalman's paper, with G(t) = I, we can re-write the model equation as:

$$\begin{bmatrix} y_{k+1} \\ y'_{k+1} \end{bmatrix} = \begin{bmatrix} 1 & \Delta T \\ 0 & 1 \end{bmatrix} \begin{bmatrix} y_k \\ y'_k \end{bmatrix} + \begin{bmatrix} 1 & 0 \\ 0 & 1 \end{bmatrix} \begin{bmatrix} \left(-\dfrac{\Delta T^2}{2}\right)(actualg_k) \\ (-\Delta T)(actualg_k) \end{bmatrix} \tag{10.7}$$

That is, after this re-arrangement the **G** is the 2-by-2 identity matrix and **u**(t) is the 2-by-1 column vector that appears as the last vector in Equation 10.7. This re-arrangement will help us set up a Q matrix that is appropriately suited for this scenario.

Of course, application of the Kalman Filter also requires the availability of measurements collected at the same sampling instants. In this case we will propose that an instrument (e.g., a laser rangefinder), would be able to provide actual measurements of the height of the falling object every ΔT seconds. At each sampling instant this measurement will be the only component of the vector of measurements **z**, which therefore reduces it to a scalar. If the measurement values are reported in the same units (meters) for the height as represented by the first state variable in the model, y_k, then the general measurement equation (Equation 6.2), recalled here:

$$\boldsymbol{z}(t) = \mathbf{H}(t)\boldsymbol{x}(t) \tag{10.8}$$

becomes

$$z_k = [1,0]\begin{bmatrix} y_k \\ y'_k \end{bmatrix} \tag{10.9}$$

That is,

$$H = [1,0] \tag{10.10}$$

Because we have a single measurement variable, its uncertainty is given by the variance of the measurements, which is a scalar in this case:

$$\mathbf{R} = \sigma_z^2 \tag{10.11}$$

This completes the model of the scenario and the generic setup for the use of the Kalman Filter here.

10.2 STRUCTURE OF THE MATLAB® SIMULATION FOR THIS CASE

Just as it was done in the previous chapter, we will structure the MATLAB® simulation in three tiers using the functions: papersim, paperloop and onedkf. The functions papersim and paperloop will play roles that are equivalent to the roles played by gizmosim and gizmoloop in the previous chapter. That is, paperloop implements a timing loop, controlled by the discrete-time variable "`t`." In every execution of this loop, the correct set of input values for the measurement variable, (which in this case is simply the height of the falling wad of paper as measured by a range-finder), is fetched from a vector where these "measured values" have been pre-calculated. Likewise, paperloop fetches the 2-by-1 vector **u**(t) from a previously computed matrix, created by papersim.

Function paperloop receives the values of the matrices **Q** and **R**, which are fixed. It also receives the value of the sampling interval at which the model will be considered, ΔT. The total number of iterations, "`iter`," is also passed to paperloop. As always, initial values for the state vector, $\mathbf{x}_0$, and for the state covariance matrix, $\mathbf{P}_0$, are needed and they are provided to papersim.

With all the arguments received, paperloop creates **F** and **G** (which will not change) and establishes matrix **H**. Within the timing loop paperloop calls onedkf for each iteration of the Kalman Filter, writing each pair of iteration results $\mathbf{x}_A$ and $\mathbf{P}_A$ on the variables `x` and `P`, which will be used in the next call to onedkf. Further, before the end of the timing loop, there are commands that save not only $\mathbf{x}_A$ to return all the state vectors calculated to papersim, but also the two values of the Kalman Gain matrix, $\mathbf{K}_G$, and the value in the first row and first column of the matrix $\mathbf{P}_A$ calculated in each iteration. That position in $\mathbf{P}_A$ is occupied by the variance associated with the first state variable, which is, in this case, our primary variable of interest: the estimated height of the paper wad. The three matrices that are updated at the end of each Kalman Filter iteration are passed to papersim, which, in turn will return them to the workspace when the simulation function is executed.

The highest-level function that implements the simulation, papersim, receives the parameters that a user may want to change from one simulation to another, such as the assumed parameters for the "diminished downward acceleration," whose mean value is set by subtracting g_{back} from 9.81 m/s^2 and its standard deviation, stored in the MATLAB® variable `gsd`. Matrix **R** is also received as an argument for the simulation, as is the initial state variable vector, $\mathbf{x}_0$, and its initial covariance matrix, $\mathbf{P}_0$. The sampling interval to use for the simulation, ΔT, the "true initial height" of the fall, `y0tr`, and the total number of iterations of the Kalman Filter to simulate, `iter`, are also arguments of the function papersim.

Function papersim creates the external error covariance, **Q**(t), which will be fixed, as the covariance matrix for the **u**(t), 2-by-1 vector that appears as the last vector to the right in Equation 10.7. To express all four values in **Q** as a function of the variance of `actualg`, $\sigma^2_{actualg}$, we will recall the following property of the covariance function:

$$COV(ax_1, bx_2) = abCOV(x_1, x_2) \tag{10.12}$$

Accordingly, for our vector **u**, given by:

$$\boldsymbol{u}_k = \begin{bmatrix} \left(-\frac{\Delta T^2}{2}\right)(actualg_k) \\ (-\Delta T)(actualg_k) \end{bmatrix} \tag{10.13}$$

the corresponding 2-by-2 covariance matrix will be (by the definition in Equation 2.8):

$$\mathbf{Q} = \begin{bmatrix} \left(\frac{\Delta T^4}{4}\right)\sigma^2_{actualg} & \left(\frac{\Delta T^3}{2}\right)\sigma^2_{actualg} \\ \left(\frac{\Delta T^3}{2}\right)\sigma^2_{actualg} & (\Delta T^2)\sigma^2_{actualg} \end{bmatrix} \tag{10.14}$$

Therefore, papersim can create a suitable matrix **Q**, on the basis of the sampling interval, ΔT, and the variance of the fluctuations in `actualg`, $\sigma^2_{actualg}$.

For this example, papersim first creates all the necessary time series that will be accessed by onedkf and then, after the invocation of paperloop, performs some basic visualization of the results.

The computation of a vector containing `iter` samples of the "true height" of the falling paper wad is achieved by iterating over the basic free fall model (Equation 10.7), where the values of the "diminished downward acceleration," are fetched from the previously created `actualg` sequence. This sequence was generated by adding noise samples from a Normal distribution with 0 mean and standard deviation `gsd`, to the mean value, which is g-g_{back}. The values used as the **u** vector will be created from the values of `actualg`, according to Equation 10.13. Then, assigning as true initial height `y0tr` and as initial speed 0, the "true" values of height and speed are calculated for `iter` iterations. All the "true values" of height are collected in a vector called `ytr`.

The values in the vector of heights that the rangefinder would report, z, are calculated by adding to the "true height" vector samples from a Normal distribution with 0 mean and a standard deviation which is the square root of (the scalar) R. (In this example R only contains the variance of the rangefinder measurements.) This value of R has to be assigned as a proposed value for each run of the simulation.

Here we see, again, the dual impact of the uncertainty terms in the scenario. For example, the variability of the "diminished downward acceleration" is represented in the model by matrix **Q**, and it has been implemented in the creation of the "true heights" by the superposition of samples from a random distribution (with standard deviation gsd) to the mean value g - g_{back}, during the creation of the time series ytr.

After paperloop is called, the matrices returned, XAVECT, PAVECT and KGVECT, can be used for some basic display of the results. In a first graph papersim displays the "true heights" available in ytr, superimposed with the time series of heights estimated (as the first state variable) by the Kalman Filter. Further, since the scaling factor between the rangefinder measurements in z and the estimated heights is 1, it is possible to also superimpose in this first graph the rangefinder measurements, for comparison.

In a second graph, papersim plots the evolution of the element in the first row and first column of PA, which is the variance of the height estimates that the Kalman Filter produces as the first element of the posterior state vector, xA. This graph is important because we would like to see that the uncertainty of the height estimates produced by the Kalman Filter is reduced as the filter works. A third graph shows the evolution of the first element of the KG matrix, for further analysis that will be presented in the next section.

The listings for functions paperloop and papersim are presented next.

```
%%% MATLAB CODE 10.01 ++++++++++++++++++++++++++++++++
% paperloop.m—timing loop for simulation of Kalman
% Filter applied to falling motion for the estimation
% of instantaneous height and speed for the hypothetical
% falling paper wad as discussed in Chapter 5.
%
% SYNTAX: [XAVECT,PAVECT,KGVECT] =
% paperloop(zvect,u,y0,P0,Q,R,DT,iter);

function[XAVECT,PAVECT,KGVECT]=paperloop(zvect,u,y0,P0
,Q,R,DT,iter);
```

```
F = [ 1 , DT ; 0 , 1 ];
G = eye(2);
H = [1 , 0];

x = y0;
P = P0;
% Measurement time series
z = zvect;

% Set up vectors to store selected elements of xA and PA
% from all iterations
PAVECT = zeros(1, iter); %we will only store the
% variance of yk fr0m PA
XAVECT = zeros(2,iter); %we will store both yk and y'k

KGVECT = zeros(2, iter); %we will store both values in
% KG, which will be 2x1 vector

for t = 1:iter %%% ----- START OF TIMING LOOP
 [PA, xA, KG] = onedkf(F,G,Q,H,R,P,x,u(:,t),z(t));
  %[PA, xA, KG] = onedkfkg0(F,G,Q,H,R,P,x,u(:,t),z(t));

 PAVECT(t) = PA(1,1); % we are only storing the variance
 % of the first state variable
             % which is yk, located in cell (1,1) of PA
 XAVECT(:, t) = xA; % we will store the estimates of
 % both state variables contained in xA
 KGVECT(:,t) = KG;

 % pass results as inputs to the NEXT iteration:
 P = PA;
 x = xA;
end    %%% ----- END OF TIMING LOOP

end
%%% MATLAB CODE 10.01 ++++++++++++++++++++++++++++++++

%%% MATLAB CODE 10.02 ++++++++++++++++++++++++++++++++
% papersim—function to simulate the fall of a paper wad
% taking into account variable air resistance and
% implementing Kalman Filter to obtain height estimates
%
```

```
% SYNTAX:[XAVECT,PAVECT,KGVECT] =
% papersim(gback,gsd,y0tr,x0,P0,R,DT,iter);

function [XAVECT,PAVECT,KGVECT]
=papersim(gback,gsd,y0tr,x0,P0,R,DT,iter);
rng(12345,'v5normal'); % Resets the Random Number
% Generator to ver5. Normal, seed =12345;

% Calculate true heights with variable air friction
g = 9.81;

gsd2 = gsd^2; %variance of the fluctuations in actualg
DT2 = DT ^2;
DT3 = DT ^3;
DT4 = DT ^4;

% Creating matrix Q according to Equation 10.14
Q =[(gsd2 * DT4 /4),(gsd2 * DT3 /2);(gsd2 * DT3
/2),(gsd2 * DT2)];

noiseg = randn(1,iter) * gsd; % creating the
% fluctuations for actualg
actualg = (ones(1,iter) * (g—gback) ) + noiseg;

ytr = zeros(1,iter);
% Create the "true" time series of heights ytr

F = [ 1 , DT ; 0 , 1 ];
G = eye(2);
% H = [1 , 0];

% create u(t) in advance
u11coeff = DT2 / (-2);
u21coeff = (-1) * DT;
u = zeros(2,iter);
for t = 1:iter
  u(:,t)=[(u11coeff * actualg(t));(u21coeff * actualg(t))];
end

% Create "TRUE" height series, iterating over Eq. 10.7
y = [y0tr ; 0];
for t = 1:iter
  ynext = F * y + G * u(:,t);
  ytr(1,t) = ynext(1,1); % preserve in vector ytr only the
```

```
 % first value in ynext, which is the height
 y = ynext; % feed back the result in the model for
 %next iteration
end

%%%% NOW SET UP TO CALL paperloop :
% create a z time series with the laser height
% measurements, including measurement noise
mnoise = randn(1, iter) * (-sqrt(R));
z = ytr + mnoise;

% Run the timing loop
[XAVECT, PAVECT, KGVECT]=paperloop(z,u, x0, P0, Q, R,
DT, iter);

% PLOT SOME RESULTS
HeightFromKF = XAVECT(1,:);
gray6 = [0.6, 0.6, 0.6];
figure; plot(z,'Color',gray6);
hold on
plot(HeightFromKF,'k','Linewidth',1.5);
plot(ytr, 'y','Linewidth',1.5);
hold off; grid;
title('True height ,KF-estimated height and rangefinder
values');
ylabel('meters')
xlabel('Kalman Filter iterations');
legend('z','HeightFromKF','ytr','Location','southwest');

% Studying the evolution of the variance of the y
% estimate (xA(1,1)):
figure; plot(PAVECT,'k','Linewidth',1.5); grid;
title('Variance of KF-estimated height');
ylabel('squared meters');
xlabel('Kalman Filter iterations');

% Now plotting the evolution of the 1st element in KG
figure; plot(KGVECT(1,:),'k','Linewidth',1.5); grid;
title('Evolution of the first element of KG (KG1) in
this example');
xlabel('Kalman Filter iterations');

end
%%% MATLAB CODE 10.02 +++++++++++++++++++++++++++++++++
```

10.3 TESTING THE SIMULATION

Here we present the results of running a simulation of the falling paper wad scenario using function papersim. In particular, we would like to simulate the drop of a paper wad in circumstances similar to the ones Galileo would have found using the top of the Tower of Pisa as starting point. Diverse sources will indicate that the original height of the tower might be around 60 meters, with some opinions differing slightly from others. Further, is probably not easy to know how "tilted" the tower might have already been during Galileo's adulthood. In any case this gives us a good excuse to simulate a difference between the estimation of the initial height of the paper, which we will propose to be 65 meters, and the value that we will use as the "true" initial height, y0tr, which we will set as 60 meters for the simulation. Since we will assume an initial estimate of 0 m/s for the speed, the following MATLAB® commands will be used to instantiate the corresponding variables:

```
x0 = [65; 0];
y0tr = 60;
```

The initial covariance matrix for the state variable vector will be set with this command:

```
P0 = [10,0;0,0.01]; % The variance for speed is
% smaller (We are confident on 0 initial speed)
```

We will use the following values to estimate the mean level and variability of the effective "diminished downwards acceleration" time series, `actualg`:

```
gback = 0.08; % Mean value will be g-gback=
              % 9.81-0.08=9.73
gsd = 0.008; % Variability will be represented
             % by std.
             % deviation of 0.008
```

As the first state variable represents height in meters and the rangefinder measures height in meters:

```
R = 1;
```

Finally, we will simulate a sampling interval of 1 millisecond and simulate 1000 iterations, which will amount to 1 second of simulated time:

```
DT = 0.001;
iter = 1000;
```

After these assignments are made in MATLAB®, we can invoke the top-level function, i.e., papersim in this way:

```
[XAVECT,PAVECT,KGVECT] = papersim(gback,gsd,y0t
                          r,x0,P0,R,DT,iter);
```

This results in the graphs shown in Figures 10.1, 10.2 and 10.3. (In Figures 10.2 and 10.3 the data cursor has been used to show the values of the trace after 500 Kalman Filter iterations.)

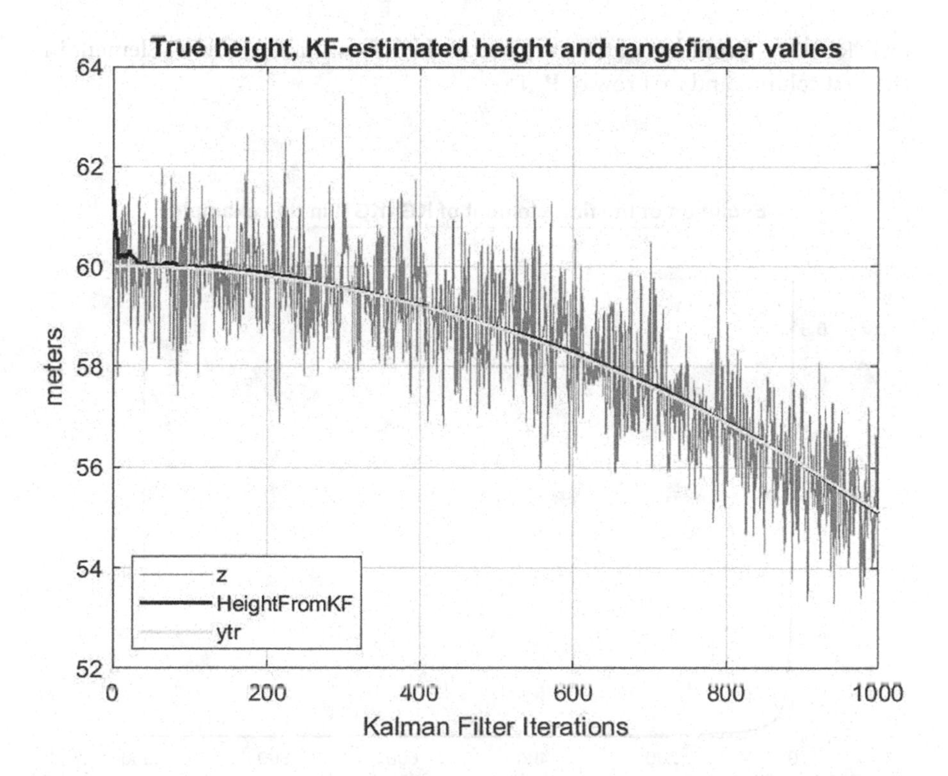

FIGURE 10.1 Plot of the simulated "true height," KF-estimated height (first element of $\mathbf{x}_A$), and rangefinder values (z) for the first 1000 iterations (the first 1 second).

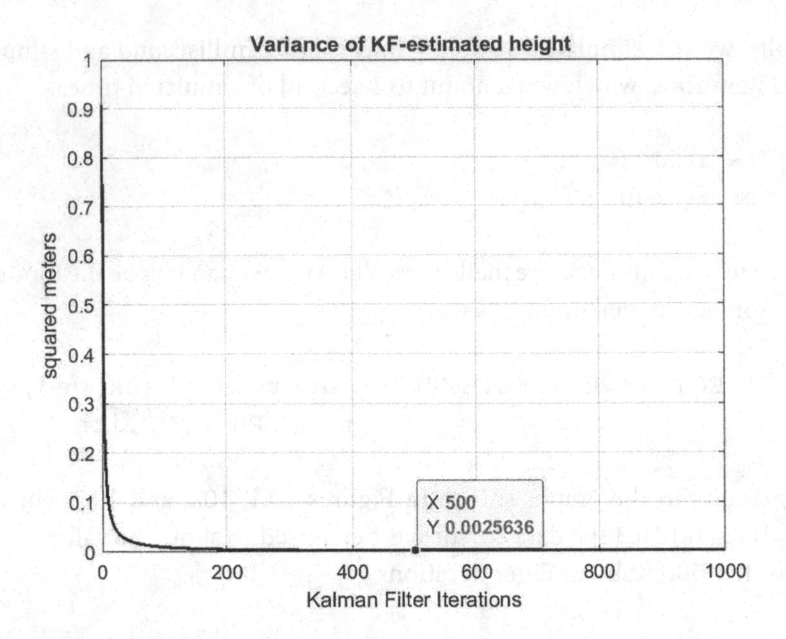

FIGURE 10.2 Evolution of the variance of the KF-estimated height (Element in the first column and first row of $\mathbf{P_A}$.)

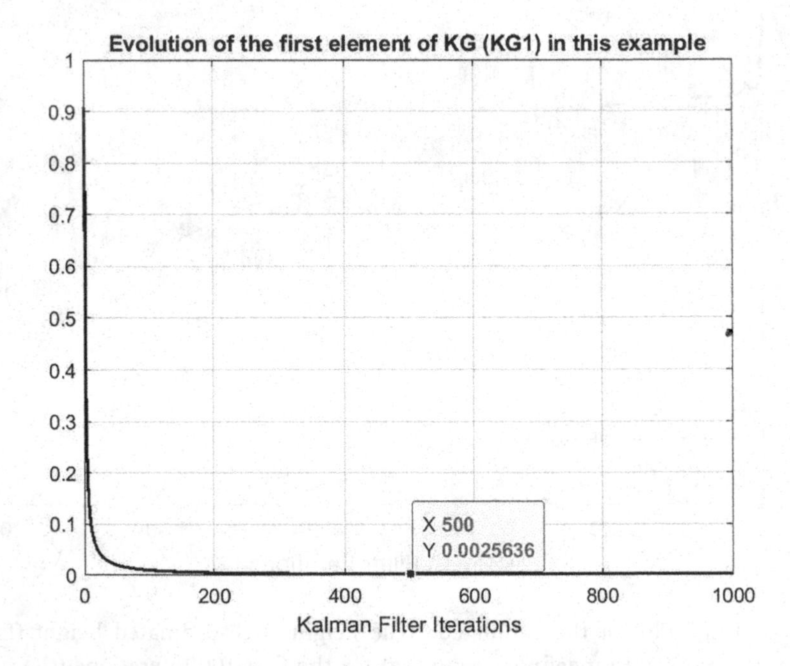

FIGURE 10.3 Evolution of the first element in vector $\mathbf{K_G}$.

Figure 10.1 shows that, while the initial estimate for the height was mistaken by 5 meters, the height estimate provided by the Kalman Filter almost instantaneously tracks the plot of true heights, approaching 60 meters in the first few iterations. This is a first evident improvement of the Kalman Filter estimate over the results that we would get from using exclusively the analytic model, without using the rangefinder measurements at all.

On the other hand, the trace displaying the rangefinder measurements is exceedingly noisy, frequently presenting values 2 meters above or below the true value. This is because we assigned a high value for R. Fortunately, the height results from the Kalman Filter are almost free from similar variations, tracking very closely the true values of height, and displaying as an almost smooth trace. This makes it evident that the height estimates from the Kalman Filter are also much better than those that could (in this particular case) be obtained from the instrumental measurements we have simulated alone. It looks like, indeed, the Kalman Filter is getting "the best of both worlds."

This is also a good opportunity to re-interpret the Kalman Filter results shown in Figure 10.1 in the typical framework of what a "filter" is expected to do, i.e., to "clean up a noisy input signal" into a "clean output signal." If we focus on the highly noisy sequence of rangefinder readings, z, as the input and the sequence of just the first element of the state variable vectors, $\mathbf{x}_A$, generated by the Kalman Filter as output, we would say that the Kalman Filter has been very successful in "cleaning the signal." The high-amplitude abrupt variations present in the rangefinder measurements have been virtually eliminated from the posterior height estimate yielded by the Kalman Filter. (This scenario was previously proposed to the reader in Figure 8.3, back in Chapter 8.)

From Figure 10.2 we appreciate that the variance associated with the Kalman Filter estimate of height decreases rapidly during the first iterations. This is a good effect, as it indicates less uncertainty associated with that estimate. Then, around iteration 500, that variance seems to stabilize, staying approximately constant.

It is also of interest to discuss the evolution of the Kalman Gain matrix elements. In this case the Kalman Gain is a 2-by-1 column vector. Further, if we recall the equation of the Correction Phase that yields the posterior state vector, $\mathbf{x}_A$ (Equation 6.38, reproduced here):

$$\boldsymbol{x}_A = \boldsymbol{x}_B + \boldsymbol{K}_G\left(\boldsymbol{z} - \boldsymbol{H}\boldsymbol{x}_B\right) \tag{10.15}$$

we notice that the correction of each of the two elements of the prior state vector are mediated separately by each of the corresponding elements of $\mathbf{K_G}$, as both are 2-by-1 column vectors and the quantity in the parenthesis is a scalar. In other words, showing the elements of all these matrices:

$$\begin{bmatrix} x_{A1} \\ x_{A2} \end{bmatrix} = \begin{bmatrix} x_{B1} \\ x_{B2} \end{bmatrix} + \begin{bmatrix} K_{G1} \\ K_{G2} \end{bmatrix} \left\{ [z] - \begin{bmatrix} 1 & 0 \end{bmatrix} \begin{bmatrix} x_{B1} \\ x_{B2} \end{bmatrix} \right\} \tag{10.16}$$

This means that the posterior *height* estimate for each iteration is found as:

$$x_{A1} = x_{B1} + K_{G1}\left(z - x_{B1}\right) \tag{10.17}$$

Where x_{B1} is the prior height estimate yielded by the model.

Therefore it is possible to visualize, in this particular case, the evolution of the "strength of the height correction" from the model result for height to the posterior estimate of height by plotting the evolution of the first element in the vector $\mathbf{K_G}$, namely K_{G1}. That is, if at a given time this first element of $\mathbf{K_G}$ were 1, the posterior height estimate would be "fully" corrected to be assigned the value measured by the rangefinder at that time. Conversely, if the first element in $\mathbf{K_G}$ were 0, then the posterior estimate of height would be assigned the value found by the model for that iteration. The evolution of that first element of $\mathbf{K_G}$ is shown in Figure 10.3. The first element of $\mathbf{K_G}$ also starts at a high value and then decreases to an approximately steady state at around 500 iterations. From this we can confirm that the Kalman Filter is capable of providing improved estimates by initially ($0 < t < 500$) using the rangefinder measurements to strongly correct the posterior height estimate (with respect to the prior height estimate, obtained from the model in each one of those iterations).

10.4 FURTHER ANALYSIS OF THE SIMULATION RESULTS

After executing papersim we will have the matrices XAVECT, PAVECT and KGVECT available in the workspace. This enables us to use alternative forms of visualization to highlight some additional points. To further emphasize the significant reduction of the uncertainty in the posterior estimate of height provided by subsequent iterations of the Kalman Filter, it is instructive to create a display of the probability density functions representing the height of the falling wad of paper that the Kalman Filter generates through its iterations. We know the characteristics (mean value and variance) of these distributions because we have access to the evolution of the first element of $\mathbf{x_A}$ and the evolution of the element in row 1 and column 1 of $\mathbf{P_A}$.

The following sequence of MATLAB® commands extracts these values from matrices XAVECT and PAVECT, and uses them to create the corresponding Gaussian distributions through the function calcgauss, which was listed in Section 7.4, as "MATLAB Code 07.02." All those Gaussian profiles are then arranged in a matrix called WATFALL, which is displayed as a MATLAB® "Waterfall plot" in Figure 10.4 and also as a top-view contour plot, in Figure 10.5:

```
%%%% MATLAB CODE 10.03 ++++++++++++++++++++++++++++++++
% Plot waterfall from height mean and variance
%
esth = XAVECT(1,:);
sdh = sqrt( PAVECT(1,:) );
szeh = length(esth);
n = linspace(0, (szeh-1), szeh );

% Plot the height estimates as waterfall plot:
hmin = 55;
hmax = 65;
hnumgauss = ((hmax–hmin) * 10) + 1;
hstep = (hmax–hmin)/(hnumgauss–1);

WATFALL = zeros(hnumgauss,szeh);
for t = 1:szeh
        [valsx,resgauss] = calcgauss(hmin,hnumgauss,
        hmax, esth(t),sdh(t));
        WATFALL(:,t) = resgauss;
end

% Create a mesh grid for waterfall contour plots:
[TIME,HEIGHT] = meshgrid(1:1:szeh, hmin:hstep:hmax);

% WATERFALL PLOT (following matlab instructions for
% "column-oriented data analysis')
figure; waterfall(TIME',HEIGHT',WATFALL');
colormap('winter'); colorbar;
xlabel('Kalman Iterations');
ylabel('Height in meters')
```

```
figure; contour3(TIME',HEIGHT',WATFALL', 50);view(2)
colormap('winter'); colorbar;
xlabel('Kalman Iterations');
ylabel('Height in meters')
%%%% MATLAB CODE 10.03 ++++++++++++++++++++++++++++++++
```

Figure 10.4 shows nicely how the Gaussian distribution for the posterior estimate of height starts up being broad (large variance), which implies considerable uncertainty, but it quickly grows and becomes thinner, so that within a few Kalman Filter iterations the corresponding Gaussian is centered very close to the "true" height for that iteration *and has now a much narrower profile (lower uncertainty).* The close approximation of the "true" height by the mean of the Gaussian distributions is better appreciated in Figure 10.5, where we are looking at the values in the WATFALL matrix from the top. After about 100 Kalman Filter iterations the peaks of the Gaussians reach higher levels, indicated by lighter-colored pixels and form a track that closely approximates the evolution of "true" heights over the first second of the falling paper wad experiment.

It is said sometimes that one of the key benefits of the Kalman Filter is that it strives to provide estimates of the state variables with small

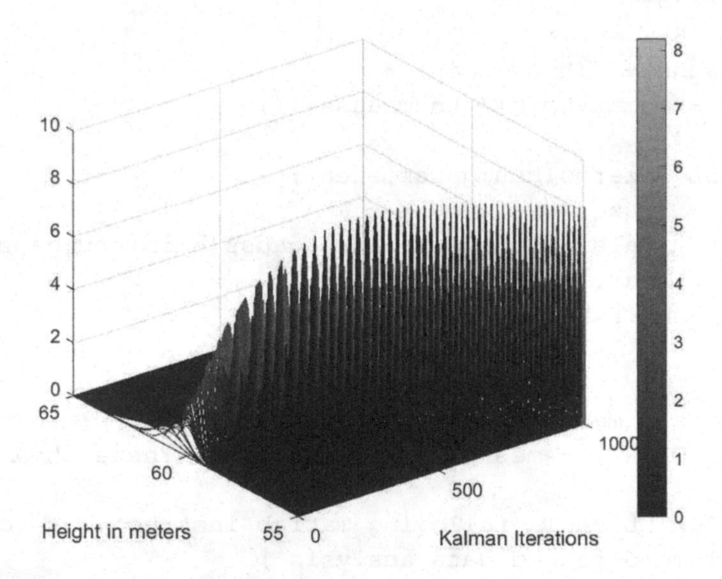

FIGURE 10.4 Sequence of Gaussian curves for the estimated height, plotted as a waterfall plot. Note how the first bell shape is broad, low and has its mean value above 60 meters, but the bell shapes become progressively thinner and taller. The means closely track the "true heights" of the simulation.

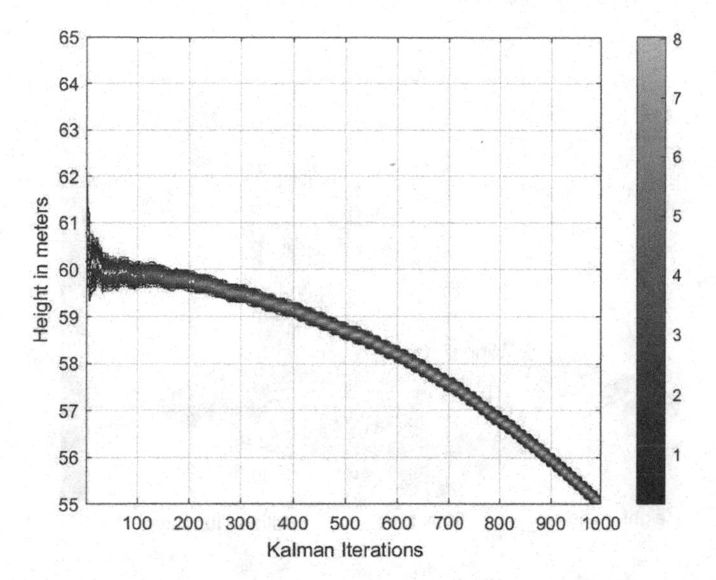

FIGURE 10.5 Sequence of Gaussian curves for the estimated height, plotted as 3-dimensional contours, seen from the top. This visualization confirms the thinning of the bell shapes and the track followed by their mean values (which is very close to the "true values" for the simulation).

variance. To further highlight how this is indeed happening with the posterior height estimates provided by our Kalman Filter, we can create one more waterfall plot, but this time involving only the first 100 iterations. The resulting plot, shown in Figure 10.6, gives us a more detailed view of the sequence of bell-shaped curves generated. The sequence of MATLAB® commands used is:

```
%%%% MATLAB CODE 10.04 ++++++++++++++++++++++++++++++
%% ZOOM INTO THE KF HEIGHT GAUSSIAN EVOLUTION FOR THE
%FIRST 100 ITERATIONS
TIME100 = TIME(:, 1:100);
HEIGHT100 = HEIGHT( :, 1:100);
WATFALL100 = WATFALL(:, 1:100);

figure; waterfall(TIME100',HEIGHT100',WATFALL100');
colormap('winter'); colorbar;
xlabel('Kalman Iterations');
ylabel('Height in meters')
%%%% MATLAB CODE 10.04 ++++++++++++++++++++++++++++++
```

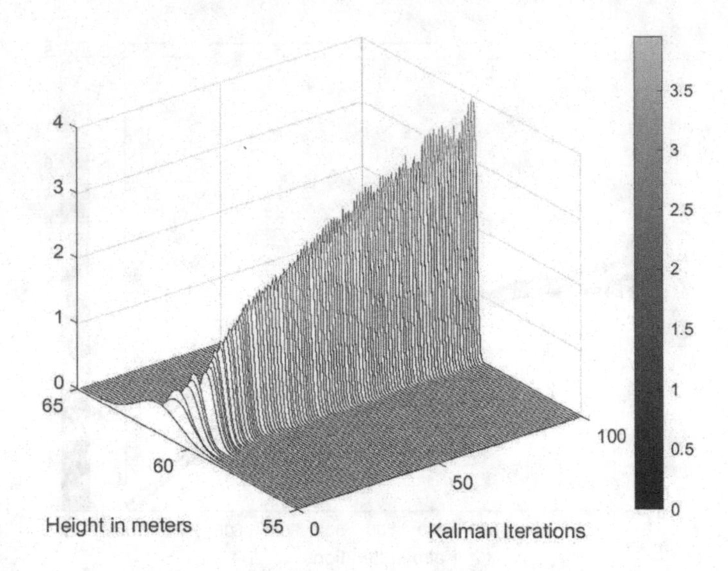

FIGURE 10.6 Sequence of Gaussian curves for the estimated height, plotted as a waterfall plot, during the first 100 iterations. This detailed view allows clearer observation of the quick thinning of the distribution for the posterior height estimate.

10.5 ISOLATING THE EFFECT OF THE MODEL

One more way in which we can dissect the results of our falling paper wad simulation is by purposely "skipping" the Correction Phase of each of the Kalman Filter iterations, to see what would happen if we only use the model (prediction) part of the Kalman Filter. In such anomalous case the overall results output by each iteration of the Kalman Filter will just be the outputs of the model ($\mathbf{x}_A = \mathbf{x}_B$ and $\mathbf{P}_A = \mathbf{P}_B$). We can easily void (bypass) the Correction Phase of the algorithm, which is the one where the information from the measurements, z, is heeded) by substituting the function onedkf in paperloop with the alternative function onedkfkg0. As explained in Chapter 9, the only difference in onedkfkg0 is that, after the correct value of KG is calculated, this function arbitrarily overwrites the complete contents of matrix KG with zeros. With this change the second terms in the correction equations (Equations 6.38 and 6.39) are cancelled and we force the outputs of the Kalman Filter iteration to be the outputs of just the model, obtained in the Prediction Phase, that is, $\mathbf{x}_A = \mathbf{x}_B$ and $\mathbf{P}_A = \mathbf{P}_B$.

If we repeat the set-up and invocation of papersim, after we replaced onedkf with onedkfkg0 in paperloop, with these commands, we obtain the results shown in Figures 10.7 and 10.8.

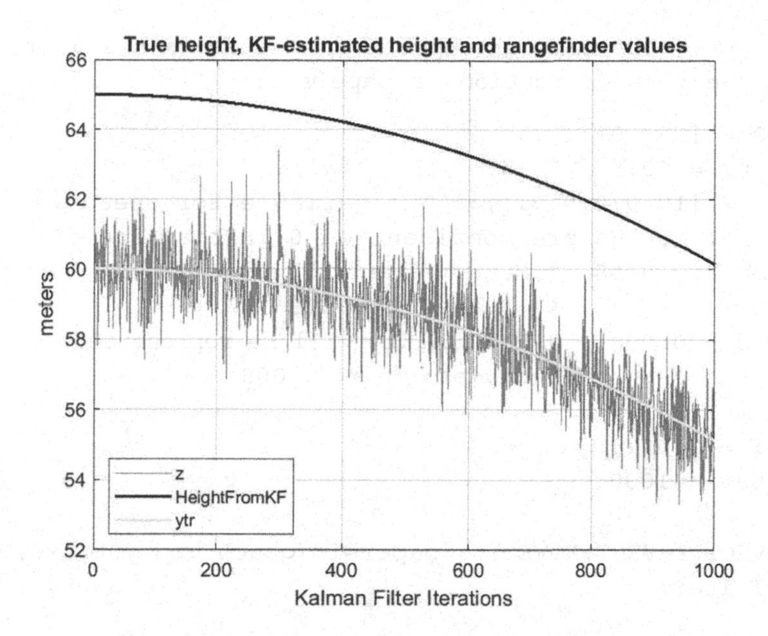

FIGURE 10.7 Plot of the simulated "true height," KF-estimated height (first element of $\mathbf{x}_A$), and rangefinder values (z) for the first 1000 iterations (the first 1 second), when the Correction Phase was anomalously bypassed.

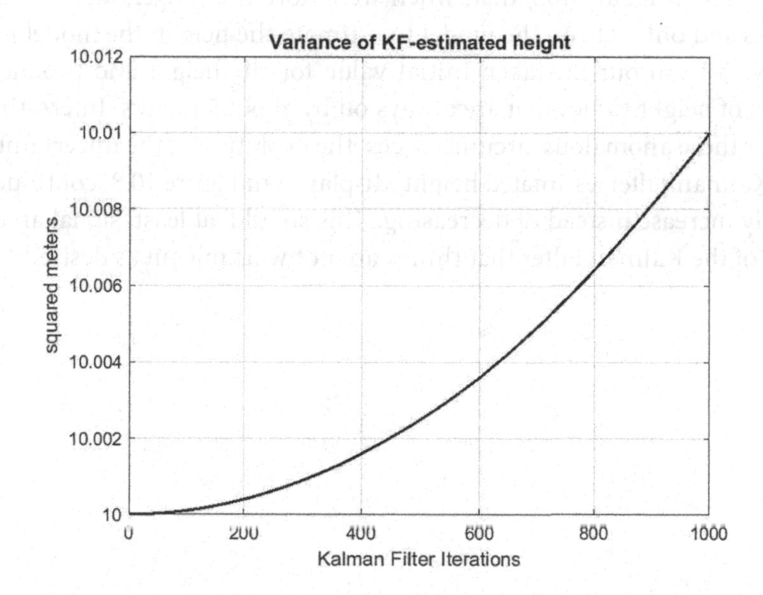

FIGURE 10.8 Evolution of the variance of the KF-estimated height (element in the first column and first row of $\mathbf{P}_A$), when the Correction Phase was anomalously bypassed.

```
% Using the same set-up and initialization as for
% the first execution of papersim:

x0 = [65; 0];
y0tr = 60;
P0 = [10,0;0,0.01]; % The variance for speed is
%smaller.(We are confident on 0 initial speed)
gback = 0.08; % Mean value will be g-gback
              % =9.81-0.08=9.73
gsd = 0.008;  % Variability will be represented by std.
              % deviation of 0.008

R = 1;
DT = 0.001;
iter = 1000;

[XAVECT,PAVECT,KGVECT] =papersim(gback,gsd,y0tr,x0,P0,
R,DT,iter);
```

This yields three plots. We will only focus on the first two of them (as the third only confirms that K_{G1} was kept constant at 0). They are shown here as Figures 10.7 and 10.8.

We see in Figure 10.7 that, when we ignore the rangefinder measurements and only rely on the model to estimate the height, the model never recovers from our mistaken initial value for the height and produces a series of height values that are always off by about 5 meters. Interestingly, under these anomalous circumstances, the evolution of the uncertainty of the Kalman Filter-estimated height, displayed in Figure 10.8, continues to slowly increase instead of decreasing. This should, at least, signal an alert user of the Kalman Filter that things are not working out as desired.

IV

Kalman Filtering Application to IMUs

IN THIS LAST PART of the book we adapt the Kalman Filtering process described and studied in previous chapters to address the problem of orientation ("attitude") estimation on the basis of signals obtained from the tri-axial accelerometer and tri-axial gyroscope contained in a Micro Electro-Mechanical System (MEMS) Inertial Measurement Unit (IMU). This is an area of application of Kalman Filtering that has attracted increased attention with the advent of MEMS IMUs which are small, cheap and energy-efficient alternatives for the monitoring and control of autonomous vehicles.

In Chapter 11 we will use our onedkf MATLAB® function (Chapter 8) to apply a Kalman Filter to the off-line estimation of orientation from signals previously recorded to a data file obtained while the IMU module was subjected to pre-scheduled changes of orientation. This, however, will require that we broaden our concept of the Kalman Filter estimation process, to accommodate a solution that uses gyroscope signals to create a state transition matrix that is different in each iteration. Therefore, in this instance, the Kalman Filter is actually "fusing" information from two types of sensors: accelerometers and gyroscopes, in a Bayesian Estimation process to yield enhanced posterior orientation estimates.

In Chapter 12 we implement a similar solution as the one implemented with MATLAB® in Chapter 11. However, in the last chapter of this book, the implementation will be developed in the C Programming Language and will run in real time, in the PC computer that hosts a MEMS IMU module.

CHAPTER 11

Kalman Filtering Applied to 2-Axis Attitude Estimation From Real IMU Signals

In this chapter we will apply the Kalman Filtering algorithm, as developed in Chapter 6 and implemented using MATLAB® in Chapter 8, to the estimation of attitude (orientation) of a commercial miniature Inertial Measurement Unit (IMU), when rotated around 2 axes. "Attitude" is the common technical term used to refer to what we intuitively perceive as the "orientation" of a rigid body. We will use both terms interchangeably in this chapter. The sensor module used was actually a Magnetic-Angular Rate-Gravity (MARG) module, as it does not only contain inertial sensors (3-axis accelerometer and 3-axis gyroscope), but it also contains a 3-axis magnetometer. However, the magnetometer outputs were not used for the Kalman Filter implementation. The accelerometer and gyroscope signals were recorded to a text file while the module was subjected to a series of pre-scheduled turns (i.e., specific changes of orientation), so that the validity and quality of the results from Kalman Filter processing can be assessed. Therefore, the application presented here is "off-line" attitude estimation. Additionally, this application of Kalman Filtering attempts to estimate attitude of an IMU module that is rotated about only 2 of the 3 orthogonal axes that describe 3-dimensional space. The listings of the MATLAB® functions discussed appear at the end of this chapter.

11.1 ADAPTING THE KALMAN FILTER FRAMEWORK TO ATTITUDE ESTIMATION FROM IMU SIGNALS

In all the previous chapters we have emphasized that the Kalman Filter merges information from a model of the dynamic system with actual (empirical) measurements of "real," observable variables that can be related to the performance of the system. So far we have proposed system models that are theoretical in nature and constant through time. The first thing we want the reader to notice about the application of Kalman Filtering presented in this chapter is that the model we will use is not a constant theoretical model. Instead, the model we will use here keeps its structure (which results from fundamental rigid body mechanics considerations), but changes the values of its parameters in every Kalman Filter iteration, based on . . . (instantaneous) measurements! Specifically, we will explain how the model for the Kalman Filter will be "customized" at every iteration by updating the elements of **F**(t), the state transition matrix, with the latest values retrieved from the 3-axis gyroscope in the IMU.

On the measurement side of the Kalman Filter, things will still be similar to previous cases. For every Kalman Filter iteration we will receive the newest values from the multi-axis accelerometer in the IMU.

So, interestingly, this application of the Kalman Filter merges information from a model and a set of measurements, but it could also be considered to be "fusing" gyroscope and accelerometer measurements towards the determination of the attitude of the IMU module (Aggarwal, Syed, and Noureldin 2010), therefore qualifying Kalman Filtering, in this application, as one of the earliest algorithms for "Sensor Fusion" (Ayman 2016).

11.2 REVIEW OF ESSENTIAL ATTITUDE CONCEPTS: FRAMES OF REFERENCE, EULER ANGLES AND QUATERNIONS

To be able to characterize attitude and, furthermore, to describe the model which will allow the "prediction" of the IMU's attitude from one sampling instant to the next, it will be necessary to adopt some of the systems available for attitude characterization. Several systems have been devised through history to describe the attitude of a rigid body and the changes of that attitude.

We should also clarify to the reader the useful interplay between the description of the orientation of a rigid body with respect to a frame of reference and our ability to characterize the change of orientation of that rigid body, from an initial orientation to a current orientation. We can describe the current orientation of the rigid body *through an "accumulated" change of orientation*, from an initial orientation (where the rigid body is considered aligned with the known reference frame), to its current orientation. We will need this point of view for the establishment of our system model since the gyroscopes do not inform us about the current orientation of the IMU. Instead the gyroscopes only provide information about small, incremental changes of orientation. It will be up to our system equations to accrue those small changes into an "accumulated" orientation change that describes, at any sampling instant, how the rigid body's orientation is changed from its initial orientation, which was the same orientation of the reference frame.

Here we will not address the formal definition of the attitude characterization systems, as that is not the topic of this book. For readers interested in studying the formal framework of those systems we can mention as a helpful reference the book by Kuipers (2002). Instead, for the limited purpose of explaining the basics of these attitude referencing systems we will use a practical analogy.

Imagine you are driving your car on a long stretch of highway that runs from South to North through a desert. Without considering the curvature of the Earth for the study of the (relatively small) motions we will discuss, we will assume a flat, 2-dimensional plane that extends according to the South-NORTH axis and the East-WEST axis, which intersect at the initial position of your car. Through their intersection we can also consider imagining a "vertical" ground-SKY axis. This set of 3 orthogonal, axes will be considered fixed throughout our discussion and will be referred to as the "Flat Earth Reference System," or the "Inertial Frame." (Please note that our discussion purposely disregards the movement of the origin point due to the rotation of the Earth and the movement of our planet around the Sun. For the purpose of this discussion we are adopting the convention that the whole desert we are traversing is flat and static.)

Now imagine you use a 5-inch string to hang an ornament (e.g., a couple of "fuzzy dice") from the rear-view mirror, or better yet, from the point of the junction between the roof of your car's cabin and its windshield that is closest to the rear-view mirror. We are going to make that point of attachment the origin of a set of axes *that move along with your car*. This is the "body frame." Imagine that, while in the driver's seat, you use a crayon to draw a line on the roof of your car's cabin in the direction of travel of your car, i.e., a line that goes from the back to the FRONT of your car, passing through the body frame origin. We will call this line the FRONT axis. Similarly, imagine you now draw a line in your roof from the right side of your car to the LEFT side of your car, also through the origin of the body frame. Finally imagine an axis, also passing through the body origin, which runs from the bottom of your car ("undercarriage") towards the ROOF of your car. That will complete the set of orthogonal "Body" axes (FRONT, LEFT and ROOF) that will always be affixed to your car (and therefore will move and turn as the car moves and turns). For the explanations in this chapter, we will assume that the IMU is affixed to the origin of the body frame (Titterton and Weston 2004), and that its axes are all aligned with the body frame just described. Under those conditions the attitude of the IMU is the attitude of your car.

In the context provided earlier, the challenge of attitude estimation is to describe the variable attitude (orientation) of your car, with respect to the fixed inertial frame. Figure 11.1 shows the relationship between the inertial frame and the body frame of reference. A key point is that the body frame of reference is attached and moves with the rigid body that we are studying (your car = the IMU module), whereas the inertial frame does not change. The attitude estimation challenge is to provide an estimate of the orientation of the rigid body, which is the orientation of the body frame, with respect to the inertial frame, for all time values.

How can we specify, quantitatively, the attitude of your car, with respect to the inertial frame? Several systems have been devised. We will only

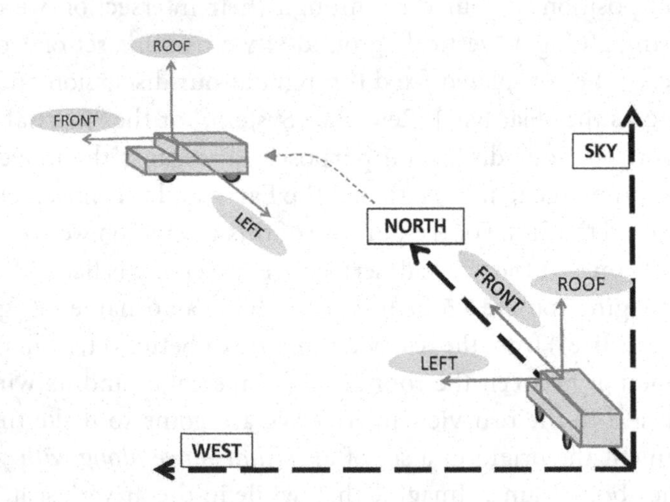

FIGURE 11.1 (Fixed) Inertial and body frames of reference. The NORTH, WEST and SKY axes define the inertial frame. The FRONT, LEFT and ROOF axes define the body frame.

briefly mention two of them. The first is based on the use of "Euler Angles" to characterize rotation (change of attitude). This system describes rotations using three angles: Phi (ϕ), Theta (θ) and Psi (ψ), which are (more or less) easy to visualize. However, representing rotations (i.e., changes of attitude) with this system has proven not to the best option for computer calculations. Instead, an alternative mathematical entity, called a "quaternion" has been adopted by many as the basis to describe changes of attitude and is in widespread use in robotics and in 3-D computer graphics. Here we will need to use both. Therefore, a very brief description of them is needed.

A succinct description of the meaning of the Euler Angles is found in the paper by Yun et al. (Yun, Bachmann, and McGhee 2008), where we have added the identification of our own practical axes, such as this: <<ROOF>>:

> It is known that a rigid body can be placed in an arbitrary orientation by first rotating it about its z-axis <<ROOF>> by an angle ψ (azimuth or yaw rotation), then about its y-axis <<LEFT>> by an angle θ (elevation or pitch rotation), and finally about its x-axis <<FRONT>> by angle ϕ (bank or roll rotation).

That is, in our scenario, if you start traveling in your car on a flat section of the desert, heading North, all 3 axes of your body frame (FRONT, LEFT, ROOF) will be aligned with (i.e., parallel to) the 3 axes of the inertial frame (NORTH, WEST, SKY). So, for as long as you move forward without turning, all three Euler Angles will be 0 degrees. However, if you encounter a hill in your path, while your car is ascending the hill your orientation will be changed so that now θ is a certain non-zero angle, determined by how steep the ascent is. If it is a light slope, θ will be small (e.g., 5 degrees). If the ascent is very steep, θ will be larger (e.g., 30 degrees). A common convention to assign the "sign" (+ or -) of the angle about a given axis will consider the angle of rotation "positive" when it is in the direction of the index, middle, ring and pinky fingers of the right hand when the thumb is extended and oriented parallel to the (positive) direction of the axis. According to this convention the θ values for the situations described would be negative. (We should warn that some texts use the "right-hand convention" for the sign of the angles, as just described, but other texts may use the opposite, "left-hand convention," instead.)

Similarly, if you need to stop for a moment and park your car on the side of the road, with its two left wheels still on the asphalt while its two right wheels end up resting on the dirt next to the road, and if the asphalt is higher than the dirt, then your car will be tilted so that now ϕ will acquire a non-zero value.

While the description of orientation change through Euler Angles implies a sequence of 3 partial rotations to characterize the complete orientation change, the characterization of a rotation using a quaternion can be visualized as a unique rotational movement. However, in this case the "axis of rotation" is not necessarily any of the 3 orthogonal axes that form part of the frame of reference. Instead, the description of each rotation is specified around a "custom" axis of rotation, specified by a 3-dimensional vector (Yun, Bachmann, and McGhee 2008):

$$\boldsymbol{u} = \begin{bmatrix} u_1 \\ u_2 \\ u_3 \end{bmatrix} \tag{11.1}$$

where u_1, u_2 and u_3 are the coordinates of the custom axis of rotation in the frame of reference. In addition the specific amount of rotation about this custom axis, which we can call β, must be specified. Both the custom axis of rotation and the amount of rotation will be encoded in 4 real numbers, forming a quaternion: q_0, q_1, q_2 and q_3. In addition, to be an actual "unit quaternion" which can be used to characterize a rotation, the following relationships must be verified (Yun, Bachmann, and McGhee 2008):

$$q_0 = \cos\left(\frac{\beta}{2}\right) \tag{11.2}$$

$$\begin{bmatrix} q_1 \\ q_2 \\ q_3 \end{bmatrix} = \boldsymbol{u}\, sin\left(\frac{\beta}{2}\right) = \begin{bmatrix} u_1 \\ u_2 \\ u_3 \end{bmatrix} \sin\left(\frac{\beta}{2}\right) \tag{11.3}$$

$$q_0^2 + q_1^2 + q_2^2 + q_3^2 = 1 \tag{11.4}$$

The four numbers, q_0, q_1, q_2 and q_3, are commonly handled together by writing them as the four elements of a 4-by-1 matrix, that is, a column vector with four elements, for example:

$$\boldsymbol{q} = \begin{bmatrix} q_0 \\ q_1 \\ q_2 \\ q_3 \end{bmatrix} \tag{11.5}$$

In our discussion, when we refer to a quaternion we will present the corresponding variable in bold typeface, as we have done for matrices and vectors. In a quaternion q_0 is commonly called the "scalar part" of the quaternion and $[q_1, q_2, q_3]$ are grouped as a 3-dimesional vector and called the "vector part" of the quaternion. *When writing the four elements in a column vector some authors place the scalar part, q_0, at the top, while some other authors may place it at the bottom. This may be a source of confusion.* In addition to the conditions expressed by Equations 11.2 to 11.4, quaternions have associated with them a number of operations, such as addition and multiplication. While quaternion addition is very similar to the standard matrix addition, quaternion multiplication is defined in a way that is different to matrix multiplication, which is not defined for two matrices when both have 4-by-1 dimensions. In fact, the analytical rotation of a 3-dimensional vector on the basis of quaternions requires a double quaternion product involving the original 3-dimensional vector ("augmented" to become a four-element quaternion), the quaternion that characterizes the rotation and its "quaternion conjugate." These developments will not be covered here. The interested reader is directed to the book by Kuipers (Kuipers 2002) for all those details. Here our objective has been just to present to the reader what a quaternion is, and how it will "look" in an analytical expression, so that we can present the kinematic equation, as the basis for the model in our Kalman Filter, in terms of quaternions.

11.3 CAN THE SIGNALS FROM A GYROSCOPE BE USED TO INDICATE THE CURRENT ATTITUDE OF THE IMU?

The gyroscopes in a typical miniature IMU ("rate gyroscopes") will provide us with numbers (one per "axis") that reflect the speed of rotation that the IMU is experiencing about each of the axes that form the body frame of reference. These will be instantaneous approximations of the speeds of rotation. None of those three numbers represents (in itself) information about the *current* attitude of the IMU. They only inform us about the speed of rotation of the IMU approximated at one point in time. Therefore, the actual "current orientation" of the IMU requires calculation of an "accumulated" overall rotation accrued from the initial state in which the body frame was aligned with the inertial frame. A simplistic and familiar analogy is the determination of the position of a runner on a track during a 100-meter dash. If we can get an estimate of the runner's speed every second, and if we assume that during each different second interval the speed is constant, when we get the speed reading of 5 m/s after the first second of the race, we could estimate that the runner was 5 meters from the starting line. If then, after 2 seconds we get a reading of 7 m/s, we could estimate that the runner is 5 + 7 = 12 meters from the starting line. If after 3 seconds we get a speed reading of 9 m/s, our estimate for position after 3 seconds could be that the runner is 5 + 7 + 9 = 21 meters from the starting line, and so on.

The model for our Kalman Filter will express a similar change from a given sampling instant to the next, in our variable of interest, which is IMU attitude, expressed as a quaternion. The development of such discrete-time model to predict the "next accumulated quaternion" from a previous one, using knowledge of instantaneous observations of speed of rotation about the axes of the body frame is presented in the book edited by Wertz (Wertz 1980). The model derived there indicates:

$$\boldsymbol{q}(t+\Delta t) \approx \left\{\boldsymbol{I}_4 + \left(\frac{\Delta t}{2}\right)\boldsymbol{\Omega}\right\}\boldsymbol{q}(t) \tag{11.6}$$

where $\boldsymbol{I}_4$ is the 4-by-4 identity matrix and the 4-by-4 matrix $\boldsymbol{\Omega}$ contains the speeds of rotation about the IMU body frame axes (measured by the gyroscopes), designated in that book as ω_u, ω_v and ω_w:

$$\boldsymbol{\Omega} = \begin{bmatrix} 0 & \omega_w & -\omega_v & \omega_u \\ -\omega_w & 0 & \omega_u & \omega_v \\ \omega_v & -\omega_u & 0 & \omega_w \\ -\omega_u & -\omega_v & -\omega_w & 0 \end{bmatrix} \tag{11.7}$$

Therefore, we will adopt this format of the kinematic equation of motion as our model in the Kalman Filter, to predict the one-step advancement of the state variables in the "accumulated orientation quaternion" from time, t, to the next time, t + 1, represented in Equation 11.6 as $\mathbf{q}(t + \Delta t)$. Accordingly, our state vector will be the four-element quaternion, $\mathbf{q}$, and the time-varying state transition matrix will be:

$$\boldsymbol{F}(t) = \boldsymbol{I}_4 + \left(\frac{\Delta t}{2}\right)\boldsymbol{\Omega} \tag{11.8}$$

11.4 CAN WE OBTAIN "MEASUREMENTS" OF ATTITUDE WITH THE ACCELEROMETERS?

The accelerometers in the IMU are not, in principle, orientation sensors. Each one of them is meant to generate an output that represents the total acceleration that the IMU is experiencing in the direction of its corresponding axis.

However, if the IMU is either static or moving in such a way that the speeds along the directions of the axes in the body frame are constant, then the only acceleration that would be sensed by the three accelerometers in the IMU will be the acceleration of gravity: g, which is approximately 9.81 m/s^2, and, most importantly, is always pointing in a direction opposite to our "SKY" axis in the inertial frame of reference.

The consistency of gravity, both in magnitude and in its alignment with respect to the inertial frame of reference is what enables the indirect approximate measurement of the IMU's attitude (with respect to the inertial frame), if the IMU is approximately in either of the states mentioned (static or moving at constant speeds in the 3 axes of the body frame).

For simplicity, let us examine the most evident case for the conversion of gravity measurements by the accelerometers to the corresponding change in one of the Euler Angles. For this, let us go back to the scenario we used to introduce the inertial frame and the body frame. In the initial condition, when both frames of reference are aligned and your car is static and facing towards the North, the ROOF direction of the body frame coincides with the SKY direction of the inertial frame, and the acceleration of gravity acts exclusively in the direction opposite to the ROOF direction of the body frame. Therefore, the accelerometer in the ROOF direction will read an acceleration of -g, while the other two accelerometers (in the FRONT and LEFT directions) will measure zero acceleration, because the angles between the direction of gravity and the FRONT and LEFT axes are 90 degrees. In this situation the ornament hanging from the origin of the body frame will stay exactly under that origin. The string and the ornament act as a plumb line and help us visualize the orientation of gravity.

Now imagine the car has advanced moving directly towards the North and it finds a hill that it must climb to continue. Let us imagine that the car is stopped in that section of highway, which has an incline of 30 degrees. As we described in a previous section, that would imply that $\theta = -30$ degrees. What will happen to the hanging ornament? The orientation of the body frame (i.e. the car) has changed, but the gravity will still align the string of the ornament in the direction opposite to the SKY axis of the referential frame, and the angle between the ornament's string and the FRONT axis of the body frame (i.e., the line going from the back to the FRONT that you drew) will now be smaller than 90 degrees, which can be visualized through the string holding the ornament. Under these circumstances, the gravity vector will now have a projection on the FRONT axis of the body frame, determined by trigonometry as in Equation 11.9, where θ will have a negative value:

$$a_{FRONT} = (g)(\sin\theta) \tag{11.9}$$

In the case of our second example, involving the tilting of the car due to uneven height of the left and right wheels, where we assumed the left wheels were higher, now the ornament's string will no longer be perpendicular to the plane formed by the "FRONT" and "LEFT" axes, because the car has been rotated a (positive) angle ϕ around the "FRONT" axis. In this case the gravity acceleration will have a projection on the LEFT axis of the body frame which can be determined as:

$$a_{LEFT} = (-g)(\cos\theta)(\sin\phi) \tag{11.10}$$

Notice that, in the particular case of this second example, θ is 0 degrees, and therefore its cosine is 1, making the acceleration along the LEFT direction of the body frame dependent only on the value of ϕ.

A more formal presentation of these relationships between the Euler Angles that characterize a given instantaneous orientation of the IMU and the accelerometer readings that can be expected is offered in the paper by Yun et al. (Yun, Bachmann, and McGhee 2008). In that paper, the authors use the following nomenclature for the axes of the body frame: Our FRONT axis is called "x"; our "LEFT" axis is called "y" and our "ROOF" axis is called z." The resulting relationships are:

$$a_x = (g)(\sin\theta) \tag{11.11}$$

$$a_y = (-g)(\cos\theta)(\sin\phi) \tag{11.12}$$

$$a_z = (-g)(\cos\theta)(\cos\phi) \tag{11.13}$$

If we obtain *normalized* accelerations, dividing the raw accelerations by $g = 9.81\ m/s^2$, we can derive two equations that, in sequence, would allow us to approximate the Euler Angles θ and ϕ from the (normalized) accelerometer measurements.

From Equation 11.11, after normalization of the raw acceleration measurements:

$$\theta = sin^{-1}(a_x) \tag{11.14}$$

And, once θ is known, we can find ϕ as:

$$\phi = sin^{-1}\left(\frac{-a_y}{\cos\theta}\right) \tag{11.15}$$

Rotations (and rotation components) around the "ROOF" axis of our body frame cannot be quantified through the readings of the accelerometers. This is easily demonstrated by considering that, if your car was placed on a rotating platform, like the ones that exist sometimes in the showrooms of the auto dealerships, the gravity vector will always remain completely aligned (and opposite) to our "ROOF" body axis and it will never have any projection on the "FRONT" or the "LEFT" axis. Therefore, the accelerometers in these two directions will always measure 0, while the accelerometer in the "ROOF" direction will always measure -g, regardless of the amount of "yaw" rotation that the car may be experiencing. Nonetheless, Equations 11.14 and 11.15 were appropriate for the use of IMU orientation estimation in a project which pursued the use of hand orientation as a means of computer input. If a user seated in front of a computer has an IMU attached to the dorsal surface of his/her hand, the articulation in the wrist lends itself naturally to rotations about the "LEFT" axis (θ), and around the "FRONT" axis (ϕ), but not to rotations around the "ROOF" axis. Accordingly, we tested the functionality of the IMU applying rotations that involved only variations on θ and ϕ, while attempting to keep ψ at a value of zero, throughout.

It's worthwhile to mention here that Farrell (Farrell 2008) proposes two alternative equations for the calculation of θ and ϕ from accelerometer measurements. However, his proposed equations are based on the calculation of inverse tangent functions, and the tangent function contains discontinuities (whereas sine and cosine are continuous). Therefore, we prefer the use of Equations 11.14 and 11.15.

Fortunately, the parameters that describe rotations in a given system (e.g., Euler Angles) can be mapped to the parameters used in a different system (e.g., the four quaternion elements). We use those kinds of conversions to be able to present the "measurements" for the Kalman Filter, originally obtained as Euler Angles, in quaternion form, which is the representation of our state variables. A conversion from quaternion to Euler Angles format is also used to be able to graph the evolution of the IMU orientation in a way that is easier to interpret. While we will not develop these conversion expressions in detail, they have been developed in the literature related to fields that deal with rotations, such as robotics, aeronautics or computer graphics, amongst others. In particular, Shoemake (Shoemake 1985) presents a succinct development of the conversion equations to go from the Euler Angles: Phi (ϕ), Theta (θ) and Psi (ψ), to an equivalent quaternion, under specific circumstances. He proposes to consider a sequence of 3 elementary rotations: First rotating by an angle Phi (ϕ) around the ("his") z axis, then rotating by an angle Theta (θ) around the y axis and lastly rotating by an angle Psi (ψ) around the ("his") x axis. He presents the three quaternions that are the equivalent to the 3 individual rotations and then, by multiplying them, in sequence, he obtains an expression for each one of the quaternion elements in his generic quaternion, which he presents as:

$$q = [w, (x, y, z)] \tag{11.16}$$

(That is, the "first" out of the four numbers is the "real" part of his quaternion.) With *those conventions* and *for those circumstances* the expressions he finds are:

$$w = \cos\frac{\psi}{2}\cos\frac{\theta}{2}\cos\frac{\phi}{2} + \sin\frac{\psi}{2}\sin\frac{\theta}{2}\sin\frac{\phi}{2} \tag{11.17}$$

$$x = \sin\frac{\psi}{2}\cos\frac{\theta}{2}\cos\frac{\phi}{2} - \cos\frac{\psi}{2}\sin\frac{\theta}{2}\sin\frac{\phi}{2} \tag{11.18}$$

$$y = \cos\frac{\psi}{2}\sin\frac{\theta}{2}\cos\frac{\phi}{2} + \sin\frac{\psi}{2}\cos\frac{\theta}{2}\sin\frac{\phi}{2} \tag{11.19}$$

$$z = \cos\frac{\psi}{2}\cos\frac{\theta}{2}\sin\frac{\phi}{2} - \sin\frac{\psi}{2}\sin\frac{\theta}{2}\cos\frac{\phi}{2} \tag{11.20}$$

Equations 11.17 to 11.20 display the general structure of the equations for conversion of angles Phi (ϕ), Theta (θ) and Psi (ψ) to a quaternion that represents the same rotation. However, a particular implementation, such as the one in our MATLAB® functions, *will follow specific conventions* (e.g., right-hand vs. left-hand sets of orthogonal axes; right-hand vs. left-hand sign rule for the angular rotation around a given axis; whether the "real" component of the quaternion is placed as the first or the last component; etc.). In particular, to match the conventions used during the recording of the gyroscope and accelerometer signals in the pre-recorded file that we will process, our MATLAB® function uses equations *similar* to Equations 11.17 to 11.20, *except that they are re-ordered.*

It should be noted, however, that, while the concepts and equivalences presented are of a general nature, their implementation must be carefully tailored in response to the specific conventions, nomenclature and settings used throughout the complete recording and processing procedure. Some IMUs (such as the one used in our implementation), have re-writable setting registers that can be used for altering the right-hand vs. left-hand character of the body frame of reference. Similarly, these settings can be used to reverse the "positive direction" of the axes (for example, from "positive up" to "positive down"). Further, the commands in the Application Programming Interface (API), commonly made available by the manufacturers will likely transmit the three components of the acceleration readings and the three components of the gyroscopic readings in a specific order, which may or not match the order in which a given algorithm expects to arrange the elements of the acceleration and speed of rotation vectors. In addition, the ordering of the four quaternion components in the 4-by-1 vector that will represent the quaternion in the computations must be made to match the order in which they have been assumed for a given theoretical algorithm (formula), or appropriate adjustments to the formula must be performed. In summary, given the variety of rotation characterization systems, quaternion syntaxes and implementation conventions, the reader is advised to pay close attention to these aspects of the actual implementation for this application of Kalman Filtering.

11.5 SUMMARY OF THE KALMAN FILTER IMPLEMENTATION FOR ATTITUDE ESTIMATION WITH AN IMU

As indicated, the following MATLAB® functions implement a Kalman Filter for attitude estimation of an Inertial Measurement Unit reading accelerometer and gyroscope signals that were collected and stored to a text file while the IMU was rotated in 2 axes.

The state vector for our Kalman Filter will be the 4-by-1 current orientation quaternion. Therefore, the model for our Kalman Filter will operate on the current quaternion representing the body frame orientation, to yield an updated version of that quaternion. This model will be implemented by a state transition matrix, $\mathbf{F}(t)$, updated on every Kalman Filter iteration, according to Equation 11.8, where $\mathbf{\Omega}$ will be a 4-by-4 matrix populated in a fashion *similar* to Equation 11.7, with the recently received speeds of rotation. (The arrangement of the speeds of rotation inside $\mathbf{\Omega}$ has to be adjusted to account for differences between the theoretical formulas and practical characteristics of our implementation. The adjustments account for different labeling and "polarity" of body axes and the different arrangement of the four quaternion elements within the 4-by-1 vector used for calculations.) In this case, there are no "control inputs" considered. Therefore, in all iterations this part of the model will be voided by zeroing **G**(t) and **u**(t).

Since the goal is to always have in the estimated quaternion a representation of the rotation necessary to re-orient the body frame from its initial orientation (where it is assumed to be aligned with the inertial frame) to its current orientation, the initial state vector must be $\mathbf{x}_0 = [1\ 0\ 0\ 0]^T$, which means "no rotation," to reflect that initial common alignment of both frames.

The measurement component of the Kalman Filter will be obtained by plugging in the corresponding normalized accelerometer readings in Equations 11.14 and 11.15. This will yield the current measured values of θ and ϕ. Since the scope of our implementation involves only rotations according to these two Euler Angles, ψ will be assigned a value of 0, and the three Euler Angles will be plugged into conversion formulas that will yield a "measurement quaternion," **z**. Acknowledging that the Euler-to-quaternion conversion is performed as a previous, separate step, matrix **H** can be the identity matrix of order 4, for all iterations of the Kalman Filter.

For the assignment of values to the covariance matrices $\mathbf{P}_0$, **P**, **Q** and **R**, representing the uncertainty in the initial state vector, the uncertainty in the state vector in general, the external noise contribution to the state vector and the noise in the measurements, we do not have definite guidelines in this case. Because we do not have specific knowledge about interactions between the components of the vectors $\mathbf{x}_0$, **x** and **z**, we will default to assigning zero values to the off-diagonal elements of $\mathbf{P}_0$, **P**, **Q** and **R**. Further, we do not have evidence that any of the diagonal elements of these matrices (variances of the corresponding vector elements) should be bigger than the others. Accordingly these matrices will take the form of a 4-by-4 identity matrix multiplied by a single value in each case. That single value will be representative of how much uncertainty we believe the corresponding vector (such as $\mathbf{x}_0$, **x** or **z**) will have.

11.6 STRUCTURE OF THE MATLAB® IMPLEMENTATION OF THIS KALMAN FILTER APPLICATION

In spite of how different this application is from previous implementation examples, we will still use the same onedkf.m MATLAB® function at the core of this implementation. As in previous examples onedkf will be called within a timing loop established in function att2loop. Its listing is included at the end of this chapter. In this example, however, att2loop does more than just calling onedkf in every iteration of the loop. In this case the state transition matrix, **F**, is built for each iteration of the timing loop, which has a control variable, t. For each t, the corresponding speeds of rotation, which were recorded by the gyroscopes, are written into parameters a, b and c, and then used to populate the $\boldsymbol{\Omega}$ matrix (adjusted for the specific conditions of our implementation), which is used to create the state transition matrix for the current iteration.

Similarly, the values of θ and ϕ for this iteration are defined on the bases of the normalized accelerometer values for this iteration. According to the assumptions for the development of this example, ψ is assigned a value of 0. Then these three angles are converted to the current measurement quaternion, **z**. At this point, onedkf can be called (with **G** and **u** filled with zeros).

Still within the timing loop, the result of each iteration, namely the orientation quaternion contained in the state vector, **x**, is converted to Euler Angles, ϕ, θ, ψ and these resulting values are preserved in a matrix that stores all the filtered Euler Angles (`filtPTP`) so that the complete time evolution of these angles can be plotted after the completion of the loop. Similarly, the Euler Angles calculated from the accelerometer values are stored in a matrix named `PTP`, for plotting after the loop is completed. Finally, the orientation quaternion obtained as posterior estimation of the state vector is similarly stored in matrix `fQuat`.

Function att2loop preserves important information about the evolving uncertainty of the model (reported by onedkf in the matrix `PA`) and the Kalman Gain matrix (reported by onedkf in matrix `KG`). As we mentioned previously, since we do not have information about the covariance values between the elements of the state vector **x** and the measurement vector **z**, and because **G** and **u** are both zeroed, **R** and $\mathbf{P}_0$ are diagonal matrices in this case and, because of that, $\mathbf{P}_A$ and $\mathbf{K}_G$ will also be diagonal, for all the iterations. Accordingly, we can collect the diagonal elements of $\mathbf{P}_A$ and $\mathbf{K}_G$ in column vectors which can be written in the t column of the corresponding storage matrices `PAd` and `KGd`, which are returned by function att2loop. Similarly, all the "filtered quaternions" that have been collected in "fQuat" through the iterations of the loop will be returned by att2loop as matrix `xAm`.

In this example, the display of a four-pane graphics window is also programmed within att2loop, but outside the actual timing loop. This window summarizes the results of the Kalman Filter process, displaying the raw gyroscope signals and the contents of the matrices `PTP`, `filtPTP` and `fQuat`, where the corresponding results were stored at the end of each one of the iterations of the timing loop. We will discuss these graphs in the following sections of this chapter.

As we have done for the previous MATLAB® implementations, the complete simulation is invoked from the MATLAB® command window executing a function that sets up the parameters for the Kalman Filter and calls the timing loop function. In this case, that function is att2sim. Its listing is included at the end of this chapter. Function att2sim reads the ID number of the text data file where the raw measurements from the IMU are stored (accelerometer, gyroscope and even magnetometer 3-axis data, as well as a timestamp for each block of data). These data were written to the text file while the recording software was running and the IMU was subjected to a sequence of pre-established specific rotations. The name of the data file recorded for the purpose of testing our Kalman Filter implementation is "data129.txt." Therefore, att2sim must be invoked with the number 129 as first argument. (att2sim opens the data file and fills matrices and vectors with the data by calling the function readRecordingFile, which is also included at the end of this chapter). The second, third and fourth arguments are matrices **Q**, **R** and $\mathbf{P}_0$. The graph of results shown in Figure 11.2 was obtained performing the following assignments for these variables before calling att2sim:

```
Q = 0.0001*eye(4);
R = 0.5 * eye(4);
P0 = 0.01 * eye(4);
```

and then calling att2sim in this way:

```
[PAd,xAm,KGd] = att2sim(129, Q, R, P0);
```

The higher value assigned to the variances of the components of vector **z**, reflects the larger uncertainty that we expect will be present in the accelerometer measurements (which will propagate to the four elements of the quaternion **z**, derived from them).

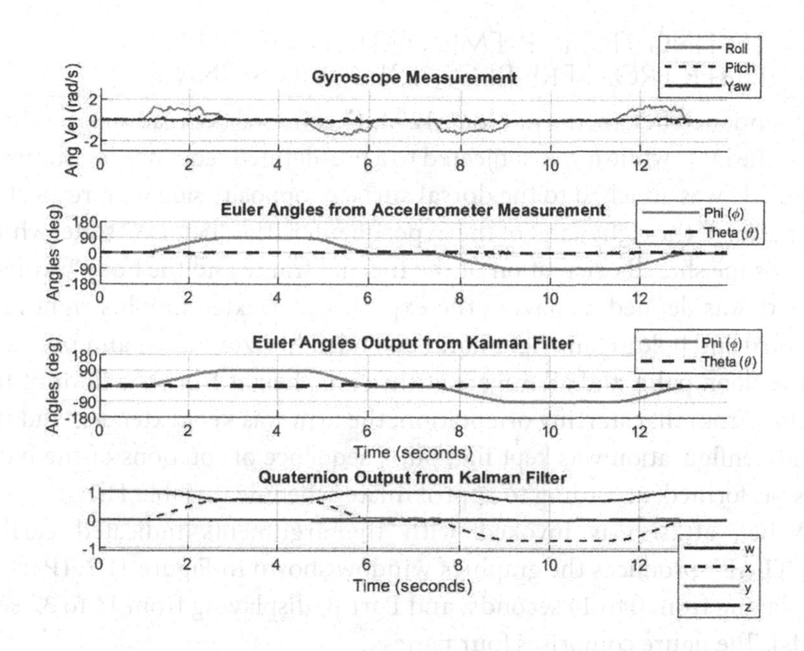

FIGURE 11.2 (Part A: 0 to 14 seconds) Graphic output from function att2sim. (The location of the legend panels was modified manually to minimize any occlusion of the traces.)

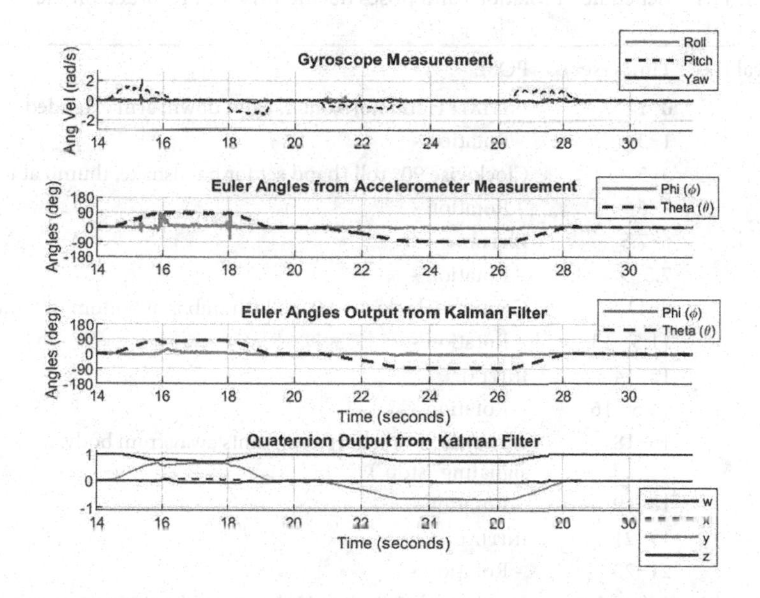

FIGURE 11.2 (Part B: 14 to 32 seconds) Graphic output from function att2sim. (The location of the legend panels was modified manually to minimize any occlusion of the traces.)

11.7 TESTING THE IMPLEMENTATION OF KALMAN FILTER FROM PRE-RECORDED IMU SIGNALS

As mentioned before, the file "data129.txt" contains the readings obtained from the IMU while it was subjected to a pre-defined sequence of rotations. The IMU was attached to the dorsal surface (opposite side with respect to the palm) of the right hand of the experimenter. The "INITIAL" state, which defines the shared orientation of the inertial frame and the body frame at time 0, was defined by having the experimenter extending his right arm forward and holding his right hand extended horizontally (palm pointing to the floor, palm and all fingers at the same height from the floor of the room). From that starting orientation, the arm was kept extended and the hand configuration was kept flat, but a sequence of rotations of the hand was performed, according to approximate schedule in Table 11.1:

When att2sim is invoked with the arguments indicated earlier, MATLAB® produces the graphics window shown in Figure 11.2. (Part A, displaying from 0 to 14 seconds, and Part B, displaying from 14 to 32 seconds). The figure comprises four panes:

TABLE 11.1 Schedule of rotations and poses during IMU test recorded on file data129.txt

Interval ID	Times (sec)	POSE
A	0–1	INITIAL: Hand horizontal, palm down, arm extended
B	1–3	- Rotation -
C	3–5	Clockwise 90° roll (hand set for handshake, thumb at top)
D	5–6	- Rotation -
E	6–7.5	INITIAL
F	7.5–9	- Rotation -
G	9–11.5	Counter-Clockwise 90° roll (thumb is at bottom of hand)
H	11.5– 13	- Rotation -
I	13–14.5	INITIAL
J	14.5– 16	- Rotation -
K	16–18	Upwards 90° pitch (palm points away from body, as if signaling "stop")
L	18–19	- Rotation -
M	19–21	INITIAL
N	21–23	- Rotation -
O	23–26.5	Downwards 90° pitch (palm of hand pointing towards body)
P	26.5–28.5	- Rotation -
Q	28.5–31	INITIAL

Figure 11.2, Pane #1 (top), contains three traces representing the raw gyroscope measurements. This shows the timing of the actual rotations applied, physically to the IMU, as highly noisy positive or negative pulses in either one of two gyroscope signals. This pane is useful in visualizing the intervals when the actual motions took place, and distinguishing the intervals when the IMU was approximately static, characterized by gyroscope signal values close to 0.

Figure 11.2, Pane #2, shows the contents of matrix "`PTP`." That is, it contains the plots for the Euler Angles θ and ϕ calculated directly from the accelerometer measurements. These sequences display some degree of noise and show a significant disturbance during the interval from 15 to 19 seconds which has a major unduly impact in the orientation estimate derived from just the measurements.

Figure 11.2, Pane #3, shows the contents of matrix "`filtPTP`," which stored the time evolution of the θ and ϕ angles obtained from conversion of the posterior quaternion estimation of the state vector. These traces, which are steady and smooth, are particularly useful in presenting the advantages gained in orientation estimation by using the Kalman Filter. Minor disturbances that are strongly reflected in the first two panes have been smoothed out here. Nonetheless, the general trends in the variations of the Euler Angles are present and comport with the patterns seen in Pane #2. The disturbance at around 16 seconds, which seems to truly be a hesitation in the orientation imposed on the IMU, is represented but does not completely disrupt the estimate of IMU orientation in the way observed in Pane #2. Similarly, the "glitch" that appears at 18 seconds on the ϕ angle derived exclusively from the "measurements" (accelerometer signals), plotted in Pane #2, has been eliminated in the trace of the ϕ angle that was obtained from the quaternion result of the Kalman Filter (Pane #3).

Lastly, Figure 11.2, Pane #4 (bottom), shows the contents of matrix "`fQuat`," displaying the time evolution of the w (scalar) and x, y and z quaternion components of the posterior estimation of the state vector. These traces are the time evolution of our actual state variables, as yielded by the Kalman Filter; however it is hard to visualize the changes of orientation of the body frame from them. (That is why the corresponding Euler Angles are displayed in Pane #3.)

It should be noted that the version of att2loop.m listed at the end of this chapter displays the full extent of the signals involved (0 to 32 seconds) in a single graphics window, which can be re-sized to full screen to look at the details, as needed. (The two separate parts, A and B, of the figure were obtained with a modified version of att2loop.m, not listed, to offer enough detail in the figure parts included in this chapter.)

Looking at the returned matrices PAd and KGd, we realize that, just like the initialization matrices used for this simulation, not only are they diagonal, but in fact they have the same value in all the entries of their main diagonals. Therefore, when we plotted them using:

```
figure; plot(transpose(PAd),'Linewidth',1.5);
grid; title('Values on the diagonal of PA');
xlabel('KF iterations')
```

and

```
figure; plot(transpose(KGd),'Linewidth',1.5);
grid; title('Values on the diagonal of KG');
xlabel('KF iterations')
```

we obtain the figures shown in Figure 11.3 and Figure 11.4. In both cases, the four traces for each figure overlap, resulting in a single visible trace.

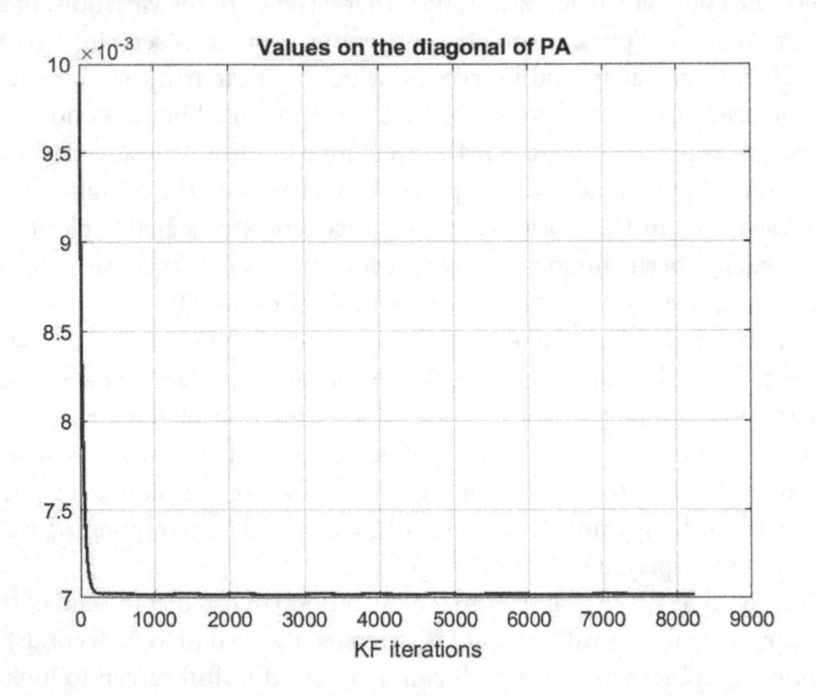

FIGURE 11.3 Evolution of the four (same) values in the diagonal of $\mathbf{P}_{\mathbf{A}}$.

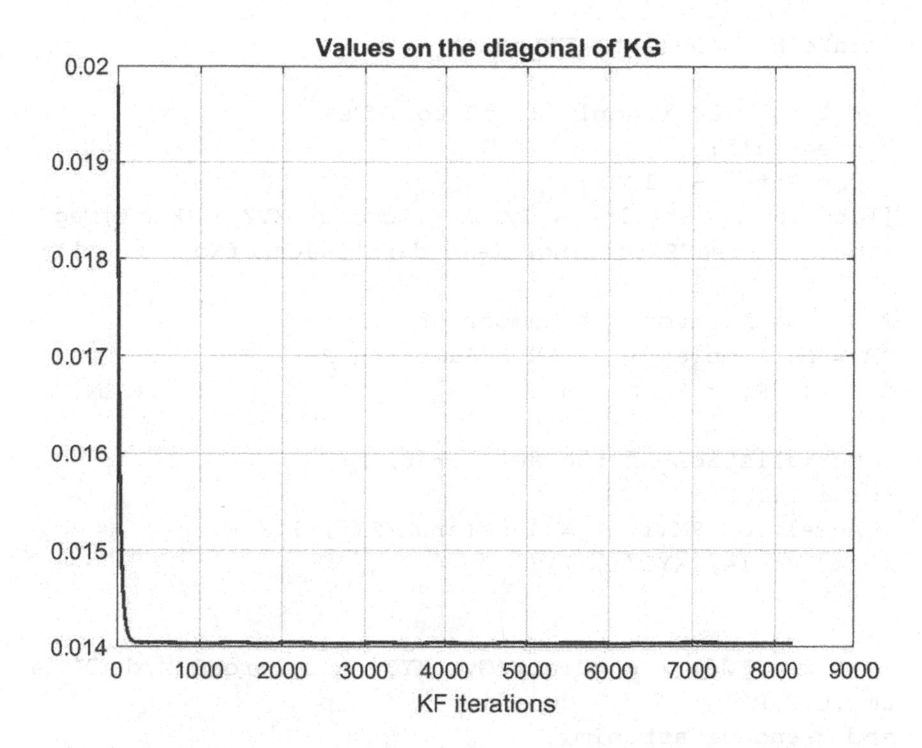

FIGURE 11.4 Evolution of the four (same) values in the diagonal of $\mathbf{K_G}$.

As we noticed in previous examples, in these cases also, the diagonal values in matrices $\mathbf{P_A}$ and $\mathbf{K_G}$ evolve from their initial values until they both reach a steady state, after about 350 iterations, indicating that the Kalman Filter has reached an equilibrium in its operation. In this case, we notice a few "bumps" in both traces, which reflect how the equilibrium is temporarily upset when rotations were applied to the IMU. After each motion, however, $\mathbf{P_A}$ and $\mathbf{K_G}$ seem to quickly revert to their equilibrium values.

```
%%% MATLAB CODE 11.01 ++++++++++++++++++++++++++++++++
% att2sim.m—function to simulate estimation of attitude
% from gyroscope and accelerometer measurements read from
% a file pre-recorded with an IMU.
%
% SYNTAX: [PAd,xAm,KGd] = att2sim(FileNum,Q,R,P0);
%
function [PAd,xAm,KGd] = att2sim(FileNum,Q,R,P0);
```

```
DataRefNo = num2str(FileNum);

% Kalman Filter Application to IMUs
% clear all;
% DataRefNo = '129';
[label,tstmp,Stillness,GyroXYZ,AcceleroXYZ,IMUquat,Magn
etoXYZ] = readRecordingFile(['data',DataRefNo,'.txt']);

N = length(tstmp); % Number of Samples
SR = N/tstmp(end); % SR = Sampling Rate
dt = 1/SR; % Sampling time

% Normaliztion of the accelerations
for i=1:1:N
  AcceleroNORM(i,:) = AcceleroXYZ(i,:)./
norm(AcceleroXYZ(i,:));
end

[PAd,xAm,KGd] = att2loop(GyroXYZ, AcceleroNORM,dt,N,ts
tmp,Q,R,P0);
end %-end of att2sim
%%% MATLAB CODE 11.01 +++++++++++++++++++++++++++++++++


%%% MATLAB CODE 11.02 +++++++++++++++++++++++++++++++++
% att2loop-Timing loop function to determine 2-D attitude
% using signals from the gyroscope and accelerometer of
% an Inertia Measurement Module (IMU)
%
%SYNTAX:[PAd,xAm,KGd] = %att2loop(GyroXYZ,AcceleroNORM,
% dt,N,tstmp,Q,R,P0);
%
function [PAd,xAm,KGd] = att2loop(GyroXYZ,AcceleroNORM,
dt,N,tstmp,Q,R,P0);

% Kalman Filter Initialization
 H = eye(4);
 x = [1 0 0 0]';
 P = P0;

 u = zeros(N,1); % There are no "Control Inputs"
```

```
  PAd = zeros(4,N); %Storage for all PA matrix diagonals
  xAm = zeros(4,N); %Storage for all posterior state vect.
  KGd = zeros(4,N); %Storage for all KG matrix diagonals
  % PTP: Matrix where we'll save the Phi, Theta and Psi
  % angles calculated directly from accelerometer readings
  PTP = zeros(N,3);
  % filtPTP: Matrix where we'll save the Phi, Theta and Psi
  % angles obtained from the quaternion output of the KF
  filtPTP = zeros(N,3);

for t=1:1:N %% BEGINNING OF THE TIMING LOOP
  % Reading the 3 signals from the gyroscopes
  a = -GyroXYZ(t,3);
  b = -GyroXYZ(t,1);
  c = GyroXYZ(t,2);

  OMEGA = [0 -a -b -c;
      a  0  c -b;
      b -c  0  a;
      c  b -a  0];
  F = eye(4) + (dt/2) * OMEGA;
  G = [0; 0; 0; 0];

  % Calculate Euler Angles from accelerometer measurement
  ax = AcceleroNORM(t,3);
  ay = AcceleroNORM(t,1);

  theta = asin( ax );
  phi = asin( -ay/(cos(theta)) );
  psi = 0;

  tht2 = theta / 2;
  phi2 = phi /2;
  psi2 = psi /2;

% Convert Euler Angles into Quaternion
z = [ cos(phi2)*cos(tht2)*cos(psi2) + sin(phi2)*
    sin(tht2)*sin(psi2); sin(phi2)*cos(tht2)*
    cos(psi2)—cos(phi2)*sin(tht2)*sin(psi2);
    cos(phi2)*sin(tht2)*cos(psi2) + sin(phi2)*
    cos(tht2)*sin(psi2); cos(phi2)*cos(tht2)*sin(psi2)—
    sin(phi2)*sin(tht2)*cos(psi2)];
```

```
 % KALMAN FILTER Iteration
 [PA, xA, KG] =onedkf(F,G,Q,H,R,P,x,u(t),z);
 %[PA, xA, KG] =onedkfkg0(F,G,Q,H,R,P,x,u(t),z); with KG=0

 PAd(1:4,t) = diag(PA); % stores current Diag. of PA
 KGd(1:4,t) = diag(KG); % stores current Diag. of KG

 x = xA; %The posterior estimate will be used for the rest
 % of the steps
 P = PA;

 filtPhi = atan2( 2*(x(3)*x(4) + x(1)*x(2)) ,
 1-2*(x(2)^2 + x(3)^2) );
 filtTheta = -asin( 2*(x(2)*x(4)-x(1)*x(3)) );
 filtPsi = atan2( 2*(x(2)*x(3) + x(1)*x(4)) ,
1-2*(x(3)^2 + x(4)^2) );

% storing results for display after completion of loop
PTP(t,:) = (180/pi)*[phi, theta, psi];
filtPTP(t,:) = (180/pi)*[filtPhi, filtTheta, filtPsi];

fPhi2 = filtPhi/2;
fPsi2 = filtPsi/2;
fTht2 = filtTheta/2;

fQuat(t,:)=[cos(fPhi2)*cos(fTht2)*cos(fPsi2) +
      sin(fPhi2)*sin(fTht2)*sin(fPsi2);
      sin(fPhi2)*cos(fTht2)*cos(fPsi2)-cos(fPhi2)*sin
      (fTht2)*sin(fPsi2);
      cos(fPhi2)*sin(fTht2)*cos(fPsi2) + sin(fPhi2)*
      cos(fTht2)*sin(fPsi2);
      cos(fPhi2)*cos(fTht2)*sin(fPsi2)-sin(fPhi2)*
      sin(fTht2)*cos(fPsi2)]';
end %% END OF THE TIMING LOOP

xAm = transpose(fQuat); % Save a matrix with all
% the posterior state vectors

% Plotting
gray6 = [0.6, 0.6, 0.6];
figure;
sp(1)=subplot(4,1,1); hold on;
```

```
plot(tstmp,-GyroXYZ(:,3),'Color',gray6,'Linewidth',1);
plot(tstmp,-GyroXYZ(:,1),'k--','Linewidth',1);
plot(tstmp,GyroXYZ(:,2),'r','Linewidth',1);
title('Gyroscope Measurement');
ylabel('Ang Vel(rad/s)');
axis([tstmp(1) tstmp(end) -3 3]);
set(gca,'Xtick',0:2:50);
legend('Roll','Pitch','Yaw','Location','BestOutside');
grid on;

sp(2)=subplot(4,1,2); hold on;
plot(tstmp,PTP(:,1),'Color',gray6,'Linewidth',1.5);
plot(tstmp,PTP(:,2),'k--','Linewidth',1.5);
title('Euler angles from Accelerometer Measurement');
ylabel('Angles (deg)');
axis([tstmp(1) tstmp(end) -180 180]);
set(gca,'Ytick',-180:45:180,'Xtick',0:2:50);
legend('Phi(\phi)','Theta(\theta)','Location','BestOu
tside');
grid on;

sp(3)=subplot(4,1,3); hold on;
plot(tstmp,filtPTP(:,1),'Color',gray6,'Linewid
th',1.5);
plot(tstmp,filtPTP(:,2),'k--','Linewidth',1.5);
title('Euler angles output from Kalman Filter');
xlabel('Time (seconds)');
ylabel('Angles (deg)');
axis([tstmp(1) tstmp(end) -180 180]);
set(gca,'Ytick',-180:45:180,'Xtick',0:2:50);
legend('Phi(\phi)','Theta(\theta)','Location','BestOu
tside');
grid on;

sp(4)=subplot(4,1,4); hold on;
plot(tstmp,fQuat(:,1),'b','Linewidth',1);
plot(tstmp,fQuat(:,2),'k--','Linewidth',1);
plot(tstmp,fQuat(:,3),'Color',gray6,'Linewidth',1);
plot(tstmp,fQuat(:,4),'r','Linewidth',1);
title('Quaternion output from Kalman Filter');
xlabel('Time (seconds)');
axis([tstmp(1) tstmp(end) -1.1 1.1]);
```

```
set(gca,'Xtick',0:2:50);
legend('w','x','y','z','Location','BestOutside');
grid on;

linkaxes(sp,'x');

end % end of att2lopp
%%% MATLAB CODE 11.02 ++++++++++++++++++++++++++++++++


%%% MATLAB CODE 11.03 ++++++++++++++++++++++++++++++++
% readRecordingFile - Function to open and read a
% text data file, written with the data recorded
% by the IMU, during its operation.
function [label,t,Stillness,GyroXYZ,AcceleroXYZ,IMUqua
t,MagnetoXYZ] = readRecordingFile(FILENAME)
label=FILENAME;
fileID = fopen(FILENAME);
readCell=textscan(fileID,'%f %f %f %f %f %f %f %f %f
%f %f %f %f %f %f','delimiter',',');
fclose(fileID);

t = readCell{1};
Stillness = readCell{2};
GyroXYZ = [readCell{3},readCell{4},readCell{5}];
AcceleroXYZ = [readCell{6},readCell{7},readCell{8}];
IMUquat = [readCell{9},readCell{10},readCell{11},read
Cell{12}];
MagnetoXYZ = [readCell{13},readCell{14},readCell{15}];
end % end of readRecordingFile
%%% MATLAB CODE 11.03 ++++++++++++++++++++++++++++++++
```

CHAPTER 12

Real-Time Kalman Filtering Application to Attitude Estimation From IMU Signals

In this chapter we will develop code for the same attitude determination problem from gyroscope and accelerometer signals as in the previous chapter, but, in this case, the Kalman Filter implementation will perform in real time, and not on previously recorded signals. In contrast to the three-tier structure of our previous MATLAB® programs, the real-time code, written in C, does not require the highest-level function (such as att2sim) since there is no need to create vectors of signal samples that will be read during the execution of the program. While the implementation of the Kalman Filter estimation will be executed in real time, the resulting estimates are not displayed or written to the console when they are generated, to avoid the additional use of time that those functions would take. Instead, the posterior estimate of orientation, $\mathbf{x}_A$, originally obtained as a quaternion, is converted to Euler Angles and stored in a text file, along with the Euler Angles from the "measurements" (accelerometer readings). Then, both sets of Euler Angles can be visualized after the execution of the real-time Kalman Filter implementation.

12.1 PLATFORM AND ORGANIZATION OF THE REAL-TIME KALMAN FILTER IMPLEMENTATION FOR ATTITUDE ESTIMATION

The sensor module for our real-time implementation of Kalman Filtering is the 3-Space LX Embedded™ (mounted on an Evaluation Kit printed circuit board), from Yost Labs Inc. Yost Labs indicates that the 3-Space LX Embedded is an ultra-miniature, high-precision, high-reliability, low-cost SMT Attitude and Heading Reference System (AHRS)/Inertial Measurement Unit (IMU) which includes triaxial gyroscope, accelerometer and compass (magnetometer) sensors. As in the case of our off-line Kalman Filter implementation, developed in the previous chapter (att2sim, att2loop, onedkf), we only use the signals from the gyroscopes and the accelerometers. The 3-Space LX Embedded can communicate with a host (PC) computer via USB. Yost Labs Inc. provides an Application Programming Interface (API) that contains low-level C functions to communicate with the IMU module. Further, Yost Labs Inc. provides some basic examples that use the functions of the API. In particular, the "streaming_example.c" from Yost Labs sets up the communication between the PC and the IMU for the repeated transmission (streaming) of IMU readings to the PC during a pre-defined amount of time. We will follow an approach similar to the one exemplified in "streaming_example.c" as the basis for the equivalent to the "timing loop" we used in our previous implementations.

Therefore, our real-time implementation program, named RTATT2IMU, will run for a pre-specified amount of time, repeatedly transmitting IMU readings to the host PC, via USB. Every time a new set of IMU readings is received, the following actions will be performed:

1. The gyroscope readings will be used to create the state transition matrix as $\mathbf{F} = \mathbf{I} + (\Delta T/2)\,\boldsymbol{\Omega}$ (Equation 11.8).
2. The accelerometer readings will be normalized and used to create the "measurement" orientation as Euler Angles.
3. The "measurement" orientation will be converted to the quaternion measurement $\mathbf{z}$. This completes the setup to perform one iteration of the Kalman Filtering algorithm (i.e., the functionality implemented, off-line, by our MATLAB® function onedkf).
4. Implementation of the Kalman Filter algorithm, performing the calculations indicated in the following five equations:

$$\boldsymbol{x}_B = \mathbf{F}(t)\boldsymbol{x}(t) + \mathbf{G}(t)\boldsymbol{u}(t) \tag{12.1}$$

$$\mathbf{P}_B = \mathbf{F}(t)\mathbf{P}(t)\mathbf{F}(t)^T + \mathbf{Q}(t) \tag{12.2}$$

$$K_G = P_B H^T \left(H P_B H^T + R \right)^{-1} \tag{12.3}$$

$$x_A = x_B + K_G \left(z - H x_B \right) \tag{12.4}$$

$$P_A = P_B - K_G H P_B \tag{12.5}$$

These are the same prediction and correction equations we arrived at in Chapter 6, (Equations 6.33 to 6.39) except that the "change of variable equations" (Equations 6.35 and 6.36) have been obviated by properly re-naming the result variables in Equations 6.33 and 6.34.

5. Conversion of the posterior orientation estimate quaternion, $\mathbf{x_A}$, to Euler Angles.
6. Writing the resulting orientation and the accelerometer-based "measurement orientation," as Euler Angles, to the output text file "RTATT2IMUOUT.txt." (This output file is overwritten with every successful run of the real-time program.)

12.2 SCOPE OF THE IMPLEMENTATION AND ASSUMPTIONS

The real-time implementation of the Kalman Filter presented in this chapter follows closely the off-line implementation developed in the previous chapter. In this case, also, the "measurement" component of the Kalman Filter framework is obtained from accelerometer readings (Equations 11.14 and 11.15), while the model uses a state transition matrix, **F**, created in each iteration from a matrix, $\mathbf{\Omega}$, populated with gyroscope readings received in the corresponding iteration (Equations 11.7 and 11.8). In this case, also, the Kalman Filter obtains the posterior orientation estimate for an IMU module that is rotated about only 2 of the 3 orthogonal axes that describe 3-dimensional space.

The reader might have noticed in the previous chapter that the covariance matrices that encode the uncertainty of measurements, the external factors in the model and the initial state vector (i.e., **R**, **Q** and initial **P**, respectively) are usually diagonal matrices. This is because (for this kind of dynamic system) we seldom have detailed information about the interactions between pairs of components in those vectors. Thus, as "default" we adopt the assumption of 0 correlation between pairs of vector components. Accordingly, all the off-diagonal elements in these covariance matrices are assigned values of zero.

Furthermore, even the diagonal values in these covariance matrices (which represent the variances of the vector components) may all be the same value. This may happen in a case such as the IMU orientation estimate because we seldom can say that the uncertainty of one element of the vector is definitely different from the uncertainty of another component. For example, we may not be able to expect a different uncertainty for the accelerations measured in the IMU's x axis than for the accelerations measured in the IMU's y axis. If all the variances are considered equal, then the **R**, **Q** and initial **P** matrices will actually be what we will call here "Scaled Identity" matrices. That is, they will fit the format:

$$I_{kn} = (k)(I_n) \tag{12.6}$$

For our real-time implementation we will assume that **R**, **Q** and the initial **P** matrices are scaled identity matrices of order 4. Accepting this constraint, which frequently occurs naturally for a scenario like the one we are addressing, will bring about important computational simplifications. Perhaps the most important simplification will be the fact that, if **P** is a scaled identity matrix and if **F** has this generic structure (as expected in this application):

$$\boldsymbol{F} = \begin{bmatrix} 1 & -a & -b & -c \\ a & 1 & c & -b \\ b & -c & 1 & a \\ c & b & -a & 1 \end{bmatrix} \tag{12.7}$$

then, the product $\mathbf{FPF^T}$ *is also a scaled identity matrix.*

In the context of our IMU orientation application, **F** will always have the format in Equation 12.7, because it is obtained as (recall Equation 11.8)

$$\boldsymbol{F}(t) = \boldsymbol{I}_4 + \left(\frac{\Delta t}{2}\right)\boldsymbol{\Omega} \tag{12.8}$$

where Ω is formed with the gyroscope readings ω_u, ω_v and ω_w, as (original format from Eq. 11.7, replicated here, then modified as in att2loop to fit our circumstances):

$$\boldsymbol{\Omega} = \begin{bmatrix} 0 & \omega_w & -\omega_v & \omega_u \\ -\omega_w & 0 & \omega_u & \omega_v \\ \omega_v & -\omega_u & 0 & \omega_w \\ -\omega_u & -\omega_v & -\omega_w & 0 \end{bmatrix} \tag{12.9}$$

To verify the previous assertion, consider the following: If **F** has the format in Equation 12.7, then $\mathbf{F}^T$ is:

$$F^T = \begin{bmatrix} 1 & a & b & c \\ -a & 1 & -c & b \\ -b & c & 1 & -a \\ -c & -b & a & 1 \end{bmatrix} \tag{12.10}$$

Then, the product $\mathbf{F}\,\mathbf{F}^T$ will be:

$$FF^T = \begin{bmatrix} 1+a^2+b^2+c^2 & a-a-bc+cb & b+ac-b-ac & c-ab+ab-c \\ a-a-cb+cb & 1+a^2+b^2+c^2 & ab-c+c-ab & ac+b-ac-b \\ b+ac-b-ac & ab-c+c-ab & 1+a^2+b^2+c^2 & bc-bc-a+a \\ c-ab+ab-c & ac+b-ac-b & cb-cb-a+a & 1+a^2+b^2+c^2 \end{bmatrix} \tag{12.11}$$

That is:

$$FF^T = \left(1+a^2+b^2+c^2\right)\begin{bmatrix} 1 & 0 & 0 & 0 \\ 0 & 1 & 0 & 0 \\ 0 & 0 & 1 & 0 \\ 0 & 0 & 0 & 1 \end{bmatrix} = \left(1+a^2+b^2+c^2\right)I_4 \tag{12.12}$$

And, representing the scaled identity matrix **P** as the product $(p_d)(\mathbf{I}_4)$:

$$FPF^T = F(p_d)I_4F^T = (p_d)FF^T = (p_d)\left(1+a^2+b^2+c^2\right)I_4 \tag{12.13}$$

So, $\mathbf{F}\,\mathbf{P}\,\mathbf{F}^T$ *is also a scaled identity matrix (of order 4).*

In our case, **H** is also a scaled identity matrix. In fact, since there is no scaling factor between the quaternion derived from accelerometer values and the quaternion generated by the model, **H** is a 4×4 identity matrix (i.e., a scaled identity matrix with a scaling factor of one).

The result obtained yields important conclusions. If the **Q** and **R** matrices are scaled identity matrices and the **P** matrix provided for initialization is also a scaled identity matrix, the result implies that the first $\mathbf{P_B}$ matrix calculated (in Equation 12.2) will be a scaled identity matrix. Since **H** is also an identity matrix, then $\mathbf{K_G}$ and $\mathbf{P_A}$ will also be scaled identity matrices. This implies that the same conditions will be met at the beginning of the second iteration, therefore perpetuating the characterization of $\mathbf{P_B}$, $\mathbf{K_G}$ and $\mathbf{P_A}$ as scaled identity matrices throughout the operation of the Kalman Filter.

One key consequence of these observations is that the term ($\mathbf{HP_BH^T} + \mathbf{R}$), which appears in Equation 12.3 will always be a scaled identity matrix. Equation 12.3 requires the computation of the inverse of this term. Fortunately, the inverse of a diagonal matrix can simply be obtained by creating a diagonal matrix where each element of the diagonal is the multiplicative inverse (e.g., 1/3) of the corresponding diagonal element of the original matrix (e.g., 3). This will greatly simplify our real-time implementation of Equation 12.3.

12.3 INITIALIZATION AND ASSIGNMENT OF PARAMETERS FOR THE EXECUTION

While the state transition matrix **F** will be calculated in every iteration using gyroscope readings, other matrices that are part of the Kalman Filter calculations will be assigned during initialization and considered constant thereafter. Since we will make these matrices scaled identity matrices of known dimensions we only need to provide the "factor" to be used for multiplying a standard identity matrix of the corresponding dimension for their creation within the program.

We have decided to include all those identity matrix scaling "factors" and other execution parameters for our RTATT2IMU program in an "initialization values text file," IVALS.txt. Beyond the scaling factors already mentioned, IVALS.txt specifies the sampling interval used for computation of **F**, i.e., ΔT, the number of the "comm port" that the PC will use to communicate with the IMU, via USB, and the total time that the program will run.

So, for example, if the only line in IVALS.txt reads:

```
8   0.08   1.0   0.001   0.001   0.01   10
```

The parameters that the real-time program, RTATT2IMU will load will be:

Comm. Port number to use: COM = 8
ΔT, sampling interval value used for calculations: dt = 0.08 (seconds)
Factor for H scaling: Hfact = 1.0
Factor for INITIAL P scaling: PAfact = 0.001
Factor for Q scaling: Qfact = 0.001
Factor for R scaling: Rfact = 0.01
Total execution duration: TIMERUN = 10 (seconds)

The real-time program RTATT2IMU will run in a console window (under the Windows OS). At the beginning of its execution, RTATT2IMU will read the initialization parameters from the file IVALS.txt and it will display the values read to the user, giving him/her the opportunity to accept those values and continue or terminate the execution of RTATT2IMU. This second option would allow the user to change the contents of IVALS.txt with a text editor (e.g., Notepad), and start the execution of RTATT2IMU again.

While the scaling factor for the initial P matrix is read from IVALS.txt, the initialization of the 4×1 state variable vector is performed inside RTATT2IMU. The state variable vector is always initialized to the quaternion $\mathbf{x} = [1\ 0\ 0\ 0]^T$ because this quaternion symbolizes "no rotation," which implies that the IMU orientation is initialized to the orientation that the module has when RTATT2IMU starts execution.

12.4 BUILDING (COMPILING AND LINKING) THE EXECUTABLE PROGRAM RTATT2IMU.EXE—REQUIRED FILES

We have created (built) the executable program RTATT2IMU.exe by compiling and linking the necessary files in Microsoft Visual Studio Community, Version 2019. We have attempted to minimize the number of custom files involved. Therefore, we have placed all the custom elements of the project in two files:

RTATT2IMU.c: C source program containing the main() function, in which the six actions described in Section 12.1 are programmed. This file also contains the necessary custom matrix manipulation functions.

RTATT2IMU.h: Header file associated to RTATT2IMU.c, where several definitions, functions and data types needed to use the API functions are included.

It should be noted that RTATT2IMU.h has the following include line:

```
#include "yei_threespace_basic_utils.h"
```

and, therefore the header file "yei_threespace_basic_utils.h," which is part of the API provided by Yost Labs, Inc., must be part of the project to create the executable. The API is provided by Yost Labs freely (for non-commercial use, under a customized GNU License), in their website: https://yostlabs.com/.

There is further explanation required for the file RTATT2IMU.c, given the fact that, in addition to controlling the flow of execution and providing the sequence for the six actions described in Section 12.1, we decided to write in it the declarations and definitions of several custom functions for matrix manipulation. (These declarations and definitions of functions could have been segregated to other files, but we chose to keep everything together in RTATT2IMU.c for simplicity.)

12.5 COMMENTS ON THE CUSTOM MATRIX AND VECTOR MANIPULATION FUNCTIONS

Implementation of the algorithm described by Equations 12.1 to 12.5 clearly requires the availability of functions to perform operations (additions, products, etc.) involving matrices and vectors, which are not normally included with many C compilers. There may be math libraries which may include these additional functions, but their use to obtain the functionality desired in the real-time implementation of the Kalman Filter we pursue here would require acquiring the libraries and integrating them into the project. Given the constrained focus (**R**, **Q** and initial **P** are scaled identity matrices) and small order (4×4 matrices, 3×1 vectors and 4×1 "quaternion" vectors), we decided to develop custom functions, using the data types vec3f and vec4f (structures defined in RTATT2IMU.h) to represent vectors and quaternions, and the generic 2-dimensional C arrays to represent 4×4 matrices.

The functions we developed (whose definitions are written towards the end of RTATT2IMU.c) will be outlined here by presenting their declaration lines and providing brief comments about each of them:

/* Functions to perform operations on "general" 4x4 matrices and/or 4x1 vectors */

`vec4f M44V41(float M44[4][4], vec4f V41);`

```
–Multiplies the 4x4 matrix M44 times the 4x1 vector
V41. It returns the result in a vec4f item (i.e., a 4
x 1 vector).
```

`void M44PLUSM44(float M1[4][4], float M2[4][4], float MR[4][4]);`

```
–Adds the two 4x4 matrices M1 and M2, writing the
result in matrix MR (does not return any items).
```

`void M44MINUSM44(float M1[4][4], float M2[4][4], float MR[4][4]);`

```
–Substracts 4x4 matrix M2 from 4x4 matrix M1, writing
the result in matrix MR (does not return any items).
```

```
vec4f V41PLUSV41(vec4f V1, vec4f V2);
```
–Adds the two 4x1 vectors V1 and V2. It returns the result in a vec4f item (i.e., a 4 x 1 vector).

```
vec4f V41MINUSV41(vec4f V1, vec4f V2);
```
–Substracts 4x1 vector V2 from 4x1 vector V1. It returns the result in a vec4f item (i.e., a 4 x 1 vector).

/* Functions to perform operations involving "SCALED IDENTITY" (SI) matrices */

```
void MAKESCALEDI4(float ScaleFactor, float M1[4][4]);
```
– Creates a 4x4 identity matrix and multiplies it by ScaleFactor, writing the result in M1 (does not return any items).

```
void SI4SI4(float M1[4][4], float M2[4][4], float
MR[4][4]);
```
–Performs the product M1 M2, where both are scaled identity matrices. Writes the result in MR (does not return any items).

```
void SI4SI4SI4(float M1[4][4], float M2[4][4], float
M3[4][4], float MR[4][4]);
```
–Performs the product M1 M2 M3, where these 3 matrices are scaled identity matrices. Writes the result in MR (does not return any items).

```
vec4f SI4V41(float SI[4][4], vec4f V);
```
–Performs the product SI V, where SI is a 4x4 scaled identity matrix and V is a 4x1 vector. It returns the result in a vec4f item (i.e., a 4 x 1 vector).

```
void SI4INV(float M1[4][4], float MR[4][4]);
```
–It calculates the inverse of a 4x4 scaled identity matrix M1. Writes the result in MR (does not return any items).

/* Function to calculate the triple product F P F^T, needed in Equation 12.2, specifically */

```
void FPFT(float F[4][4], float P[4][4], float MR[4][4]);
```
This function is not a general matrix operation function. It is meant to be used only to obtain the partial result for Equation 12.2 which requires the calculation of the term F P FT. It assumes that F is

a 4x4 state transition matrix that fits the format described in Equation 12.7. It also assumes that P is a 4x4 scaled identity matrix. Then it uses the result found for this specific situation in Equation 12.13, to determine the value of the diagonal elements in the result, and writes the resulting scaled identity matrix in MR (does not return any items).

/* Functions to "print" a 4x4 matrix or a 4x1 vector to the console (NEEDED ONLY FOR DEBUGGING) */

void PRNM44(float M[4][4]);
–Displays the contents of the 4x4 matrix M to the console as 4 lines with 4 numbers in each line (does not return any items).

void PRNV41(vec4f V);
–Displays the contents of the 4x1 vector V to the console as 4 lines with 1 number in each line (does not return any items).

These custom functions are used to implement the Kalman Filtering algorithm, as described by Equations 12.1 to 12.5 in the following segment of the RTATT2IMU.c file, where the implementations of the Prediction Phase (Equations 12.1 and 12.2) and the Correction Phase (Equations 12.3, 12.4 and 12.5) are identified. Some additional in-line comments are also included for further clarification of the way in which the custom functions implement the equations. The complete listings of RTATT2IMU.c and RTATT2IMU.h appear in Appendix A, at the end of this book.

// KF PROCESSING ######## (Uses generic 4x4 matrices ML, MO, MP, MT for TEMPORARY storage)

// ----- PREDICTION EQUATIONS

```
xB = M44V41(F, xA); // EQ 12.1: xB = F * XA (There is
no u vector in this application)

FPFT(F, PA, ML); // Term F * P * FT of EQ. 12.2 is
calculated and stored in ML
M44PLUSM44(ML, Q, PB); // Completing EQ. 12.2. Result
stored in PB
```

// ----- CORRECTION EQUATIONS

```
SI4SI4SI4(H, PB, H, MN); // PROD HPBHT in MN—Can use H
instead of HT b/c H is symmetric
M44PLUSM44(MN, R, MO); // (HPBHT + R) computed and
held in MO
SI4INV(MO, MS); // INVERSE OF(HPBHT + R) computed and
held in MS
SI4SI4SI4(PB, H, MS, KG); // Completing EQ. 12.3 – Can
use H instead of HT because it is symm.

xA = V41PLUSV41(xB, (M44V41(KG, (V41MINUSV41(z,
(M44V41(H, xB)))))))); // EQ. 12.4

SI4SI4SI4(KG, H, PB, MT); // The term (Kg * H * PB) is
computed. Result in MT
M44MINUSM44(PB, MT, PA); // COMPLETING EQ. 12.5.
RESULT IN PA
```

// END OF KF PROCESSING ########

12.6 INPUTS AND OUTPUTS OF THE REAL-TIME IMPLEMENTATION

When RTATT2IMU.exe is executed in a console window, it will first open IVALS.txt (expected to be in the same subdirectory as the executable) to read the initialization parameters written there. It will then display the parameters read from the file on the console and allow the user to enter "r" to run the program with those parameters or "q" to quit the execution, if those parameters are not adequate. In that second case, the user can open the file IVALS.txt with an ASCII text editor to update the parameters to different values and then run RTATT2IMU.exe again.

The first critical parameter that must be assigned correctly for a successful run is the serial communication port, "Comm Port," that will be used by the program to communicate with the IMU module. Fortunately, for the 3-Space LX Embedded IMU (and other of their IMUs) Yost Labs Inc. offers a free suite of PC programs which would allow the user to identify the "Comm Port" in which the USB cable to the IMU is plugged, interactively.

Another critical parameter is the ΔT, sampling interval, to be used in the Kalman Filter calculations. This parameter is involved, for example, in the scaling of the elements of the **Ω** matrix, which then is used to create the state transition matrix, **F**.

Since the acquisition of data is not controlled by a specifically timed interrupt, and instead samples are read from the IMU when the flow of the program goes through a section of a looping ("while") function, the latency between successive IMU data reads may be impacted by the speed of the particular PC host being used. The suggestion we can give regarding the assignment of ΔT in IVALS.txt is to perform a preliminary ("calibration") run of RTATT2IMU.exe, and read the average sampling interval that it calculates and displays to the console (once the TIMURUN seconds of execution are completed). Then a similar value can be written in IVALS.txt for the next (effective) run. The average sampling interval displayed to the console when RTATT2IMU.exe completes execution will already be impacted by the speed of the specific platform being used.

The rest of the parameters read from IVALS.txt really correspond to the specific instantiation of the Kalman Filter that will be implemented (Hfact, PAfact, Qfact Rfact), which will be assigned according to our intuition about the Kalman Filter algorithm, and the length of execution of the program (TIMERUN), in seconds.

As explained in previous sections, the computations pertaining to Equations 12.1 to 12.5 will be performed in real time, but these results will be written to the text file RTATT2IMUOUT.txt, for visualization *after the fact*. In this output file each line will contain the orientation directly derived from the accelerometer readings (i.e., the "measurement" for the Kalman Filter), expressed as three Euler Angles, followed by the posterior estimate of the orientation (i.e., the "result" of the Kalman Filter in each iteration), also expressed as three Euler Angles. Ahead of these six floating-point values a timestamp that will be between 0 and TIMERUN seconds is included (first column). The values are separated by commas. Therefore, the resulting output file can be read into Microsoft Excel or MATLAB®, taking the appropriate choices during the file opening process. Each line in RTATT2IMUOUT.txt contains the following values:

Time, ϕ_{acc}, θ_{acc}, ψ_{acc}, ϕ_{KF}, θ_{KF}, ψ_{KF}

Therefore, the columns in RTATT2IMUOUT.txt will be:

Column 1: Time (from 0 seconds to TIMERUN seconds).

Column 2: Phi Euler Angle from accelerometer measurements.

Column 3: Theta Euler Angle from accelerometer measurements.
Column 4: Psi Euler Angle from accelerometer measurements. (This is always assigned a value of 0.)
Column 5: Phi Euler Angle from the posterior quaternion estimate calculated by the Kalman Filter.
Column 6: Theta Euler Angle from the posterior quaternion estimate calculated by the Kalman Filter.
Column 7: Psi Euler Angle from the posterior quaternion estimate calculated by the Kalman Filter.

12.7 TRYING THE REAL-TIME IMPLEMENTATION OF THE KALMAN FILTER FOR ATTITUDE ESTIMATION

Here we outline two test runs that we performed with the real-time program RTATT2IMU.exe. In each of these test runs the IMU will be rotated about just one axis. Therefore, considered together, these two test runs involve the same motions as listed in Table 11.1, for the file where accelerometer and gyroscope signals had been recorded first and then processed after the fact, in MATLAB®. For both test runs we used the same initialization values in IVALS.txt:

8 0.08 1.0 0.001 0.001 0.01 10

Therefore, each test run lasted 10 seconds.

- Test run a:
  Starting with the IMU board parallel to the floor (horizontal), the IMU is first rotated for a 90-degree clockwise roll (so it stops when the PC board is perpendicular to the floor). Then it is returned to the horizontal position. Next, a 90-degree counter-clockwise roll is applied (that stops when the PC Board is perpendicular to the floor). Then the IMU is returned to the horizontal position.

- Test run b:
  Starting with the IMU board parallel to the floor (horizontal), the IMU is first rotated for a 90° upwards pitch, (raising the front edge of the PC board), stopping when the PC board is perpendicular to the floor. Then it is returned to the horizontal position. Next, a 90° downwards pitch is applied (that stops when the PC Board is perpendicular to the floor). Then the IMU is returned to the horizontal position.

It should be noted that the motions in both test runs were performed by holding the IMU PC Board with the right hand. Therefore, the motions were not completely smooth and the stopping positions were just determined when the PC Board was "approximately" perpendicular to the floor.

12.8 VISUALIZING THE RESULTS OF THE REAL-TIME PROGRAM

After *each* of the test runs, the results were collected in the text file RTATT2IMUOUT.txt. (*Remember that RTATT2IMUOUT.txt is overwritten by every successful run of RTATT2IMU.exe.*) Each line in this file contains seven floating point numbers, separated by commas, with the identity of the variables as described in Section 12.6. To verify the effect of the real-time Kalman Filter estimation it is instructive to compare the orientation estimated by the Kalman Filter (converted to Euler Angles), to the Euler Angles that are determined from just the accelerometer readings (i.e., "the measurements"). We have created the MATLAB® function "graphimures" for this purpose. This is how the function is used:

```
DATA = graphimures (filename , csela, cselb, cselc, cseld);
```

In this function the argument filename must be the output text file name "RTATT2IMUOUT.txt" and the remaining 4 numerical arguments (csela, cselb, cselc and cseld) define which 4 columns of RTATT2IMUOUT.txt will be plotted vs. time (in seconds). The data in the column indicated by each of these arguments will be plotted in different color and line style, as follows:

1. The column selected by csela will be plotted in black with solid line.
2. The column selected by cselb will be plotted in black with dash-dot line.
3. The column selected by cselc will be plotted in gray with solid line.
4. The column selected by cseld will be plotted in gray with dash-dot line.

This is the listing of the function graphimures.m:

```
%%% MATLAB CODE 12.01 ++++++++++++++++++++++++++++++++
% Function graphimures—displays 4 columns of
% the ascii output file RTATT2IMUOUT.txt
% versus the first column(time), for comparison
%
% SYNTAX: DATA = graphimures( filename , csela , cselb,
% cselc, cseld);
% The traces are assigned as follows:
% csela: black solid line; cselb: black dash-dot
% cselc: gray solid line; cseld: gray dash-dot
%
function DATA = graphimures( filename , csela , cselb,
cselc, cseld);

DATA = load(filename,'-ascii');
t = DATA(:,1);
CA = DATA(:,csela);
CB = DATA(:,cselb);
CC = DATA(:,cselc);
CD = DATA(:,cseld);
lgnds = ['T-i-m-e-';'Phi--Acc';'ThetaAcc';'Psi--
Acc';'Phi—KF';'Theta-KF';'Psi—KF'];
gray6 = [0.6 0.6 0.6];

figure;
plot(t,CA,'k','Linewidth',1.5);grid on;
hold on;
plot(t,CB,'k-.','Linewidth',1.5);
plot(t,CC,'Color',gray6,'Linewidth',1.5);
plot(t,CD,'Color',gray6,'Linestyle','-
.','Linewidth',1.5);
hold off
legend(lgnds(csela,:),lgnds(cselb,:),lgnds(cselc,:),lg
nds(cseld,:),'Location','Southeast');grid on;
ylabel('degrees'); xlabel('time in seconds')

end % end of function graphimures
%%% MATLAB CODE 12.01 ++++++++++++++++++++++++++++++++
```

It is suggested that the angles generated by the Kalman Filter estimation (e.g., columns 5 and 6) be plotted in solid line, as we are focusing on the effects of using the Kalman Filter to mutually enrich the information from measurements and model. Then, the black color lines can be used for the two versions of the Euler Angle that might be most relevant for a specific set of run results.

Accordingly, we used graphimures in this way after TEST RUN A:

```
DATA = graphimures('RTATT2IMUOUT.txt', 5,2,6,3);
```

and we used graphimures in this way after TEST RUN B:

```
DATA = graphimures('RTATT2IMUOUT.txt', 6,3,5,2);
```

These executions of graphimures generated Figures 12.1 and 12.2, respectively.

Both figures reveal that the orientation estimates obtained by the real-time Kalman Filter are less reactive to transient high frequency noises,

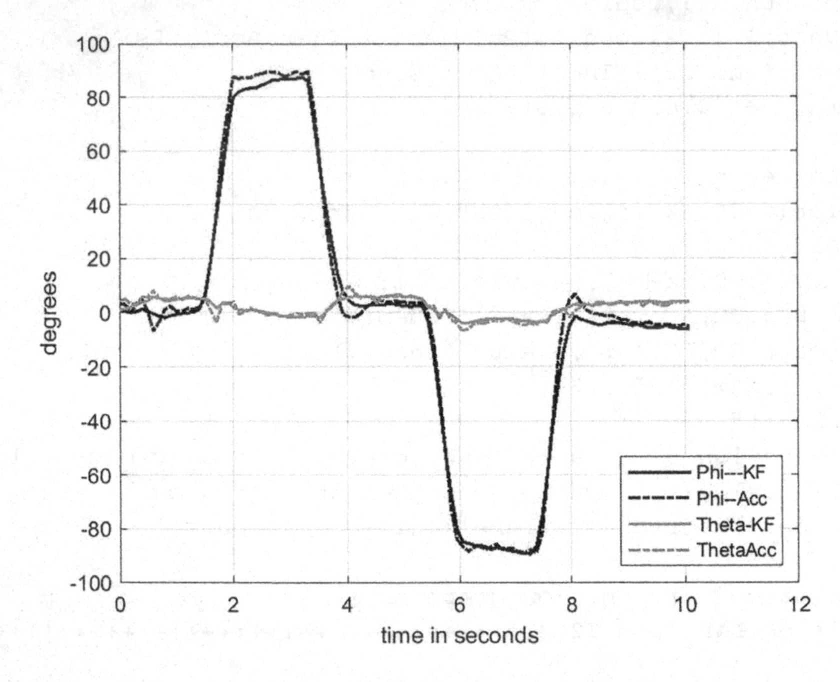

FIGURE 12.1 Comparison of the Phi Euler Angle traces generated by the accelerometer measurements (black dash-dot line), and the one that was estimated by the Kalman Filter in real time (black solid line), during TEST RUN A.

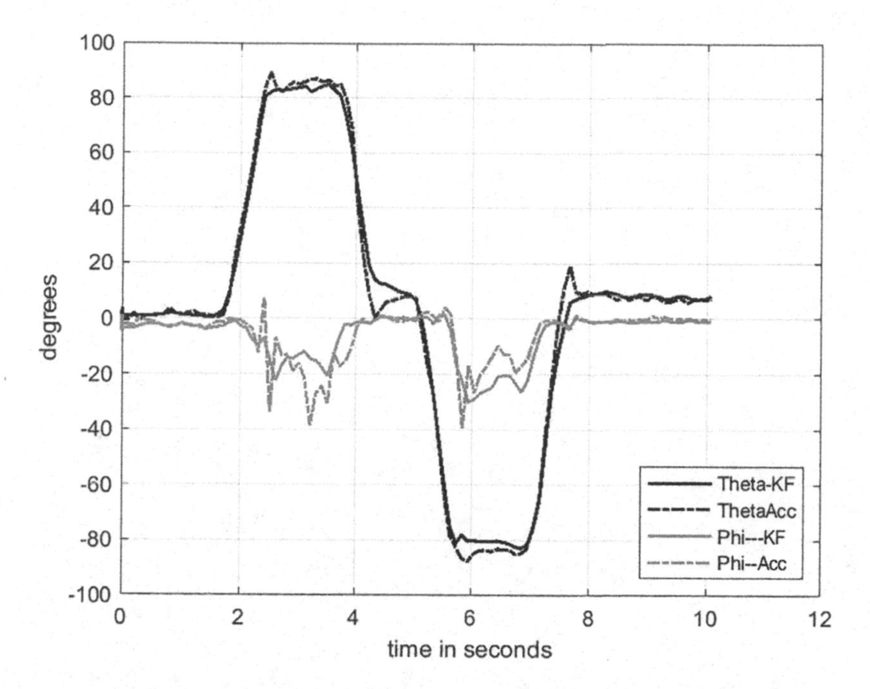

FIGURE 12.2 Comparison of the Theta Euler Angle traces generated by the accelerometer measurements (black dash-dot line), and the one that was estimated by the Kalman Filter in real time (black solid line), during TEST RUN B.

which is evidenced by the smoother solid black traces in both of these figures. This is the same kind of effect that we also observed as a result of the MATLAB® implementation of the Kalman Filter applied to pre-recorded IMU signals, in Chapter 11. In particular, Figure 12.2 shows how when the maneuver of the IMU itself was more hesitant (as indicated by the non-zero, noisy gray traces for the Phi estimates) the measurement-only estimates of the Theta angle are impacted by that noise, whereas the Kalman Filter estimates of Theta (black solid line) were much less affected by those disturbances.

APPENDIX A

Listings of the Files for Real-Time Implementation of the Kalman Filter for Attitude Estimation With Rotations in 2 Axes

A.1—RTATT2IMU.C

A.2—RTATT2IMU.H

```
// + + + + + + + FILE RTATT2IMU.c + + + + + + + + + + +
// PROGRAM RTATT2IMU.c—Implements the discrete Kalman
Filter algorithm
// for estimation of the attitude of a 3-Space Sensor
LX IMU, from Yost Labs(c),
// using the quaternion orientation as state vector
and a measurement attitude
// quaternion derived from accelerometer measurements,
under 2 assumptions:
//—Rotations are performed in only 2 axes
//—The matrices Q, H, R and the initial PA, are all
"scaled Identity" matrices
// The program runs only for a user-defined number of
seconds (TIMERUN).
// Implementation parameters (INCLUDING COMM PORT TO
BE USED) are read at run
// time from the file IVALS.txt expected in the same
directory as the executable.
// The management of the IMU data streaming is based
on the file streaming_example.c
// provided by Yost Labs (c)

#include "RTATT2IMU.h"
#include <conio.h>
#include <math.h>
#include <stdio.h>
#include <stdlib.h>
```

// Declarations for CUSTOM FUNCTIONS to implement the KALMAN FILTER

```
vec4f M44V41(float M44[4][4], vec4f V41);

void M44PLUSM44(float M1[4][4], float M2[4][4], float
MR[4][4]);
void M44MINUSM44(float M1[4][4], float M2[4][4], float
MR[4][4]);
vec4f V41PLUSV41(vec4f V1, vec4f V2);
vec4f V41MINUSV41(vec4f V1, vec4f V2);

void MAKESCALEDI4(float ScaleFactor, float M1[4][4]);
void SI4SI4(float M1[4][4], float M2[4][4], float
MR[4][4]);
```

```
void SI4SI4SI4(float M1[4][4], float M2[4][4], float
M3[4][4], float MR[4][4]);
vec4f SI4V41(float SI[4][4], vec4f V);
void SI4INV(float M1[4][4], float MR[4][4]);

void FPFT(float F[4][4], float P[4][4], float MR[4][4]);

void PRNM44(float M[4][4]);
void PRNV41(vec4f V);

float ML[4][4], MN[4][4], MO[4][4], MS[4][4], MT[4]
[4]; // Scratch-Pad matrices for intermediate results
float a, b, c;
float Acmag, AcxN, AcyN, AczN;
float tht, psi, phi;
float th2, ps2, ph2;
float flttht, fltpsi, fltphi;
float r2d = 180 / (3.1416);

int main(void)
{
 HANDLE com_handle;
 DWORD bytes_written;
 DWORD bytes_read;
 unsigned char write_stream_bytes[3];
 unsigned char write_tare_bytes[3];

 Batch_Data batch_data = { 0 };
 unsigned int sample_count; // the amount of packets
 received
 char commNum;
 char decisionChar;

 TCHAR commStr[] = TEXT("\\\\.\\COM8"); // The comm
 port # to use is overwritten from IVALS.txt

 float CurrentTime;
 FILE* fp;
 FILE* fini;

 vec3f Gy, Ac, Compass;
 vec4f IMUquat;
```

```
// THESE PARAMETERS ARE ACTUALLY READ-IN FROM IVALS.txt
  // —These are "example" values
  int TIMERUN = 10; // Duration of the streaming
  process (e.g., 10 seconds)
  float PAfact = 0.001; // Values for the main
  diagonal of INITIAL PA (e.g., 0.001)
  float Qfact = 0.001; // Values for the main diagonal
  of Q (e.g., 0.001)
  float Hfact = 1.0; // Values for the main diagonal
  of H (must be 1.0)
  float Rfact = 0.01; // Values for the main diagonal
  of R (e.g., 0.01)
  float dt = 0.06; // Values for the "Delta T" factor
  in equation 11.16 (e.g., 0.06)
  float dt2;

  // READING INITIAL VALUES FROM FILE IVALS.txt
  // EXPECTED IN SAME DIRECTORY AS EXECUTABLE
  fini = fopen("IVALS.txt", "r"); // Reads Initial values
  and parms. from this file, same directory
  rewind(fini);
  fscanf(fini, "%c %f %f %f %f %f %d", &commNum, &dt,
  &Hfact, &PAfact, &Qfact, &Rfact, &TIMERUN);
  printf("RUN PARAMETERS : COMM, dt, Hfact, PAfact,
  Qfact, Rfact, TIMERUN \n");
  printf("READ FROM IVALS.txt: %c %f %f %f %f %f %d \n",
  commNum, dt, Hfact, PAfact, Qfact, Rfact, TIMERUN);
  fclose(fini);

  vec4f xA = { 0 }; // state vector into KF (quaternion)
  vec4f xB = { 0 };
  vec4f z = { 0 };
  float F[4][4] = { 0 };
  float PA[4][4] = { 0 };
  float PB[4][4] = { 0 };
  float Q[4][4] = { 0 };
  float R[4][4] = { 0 };
  float H[4][4] = { 0 };
  float KG[4][4] = { 0 };

  float Stillness;
```

```
int i;

  sample_count = 0;

  float STREAM_DURATION = TIMERUN; //Unit SEC
  ++++++++++++ TIME ASSIGNMENT
```

// High-resolution performance counter variables. This acts as a timer

```
  LARGE_INTEGER frequency; // counts per second
  LARGE_INTEGER start_time, current_time; // counts
  QueryPerformanceFrequency(&frequency); // gets
  performance-counter frequency, in counts per second

  printf("ENTER [r] to accept parameters and run or
  [q] to quit and change IVALS.txt \n");
  decisionChar=getchar();
  printf("\nCharacter entered: ");
  putchar(decisionChar);
  printf(" \n");
  if (decisionChar == 'q')
  {
    printf("Execution will be terminated. \n");
    printf("Modify the file IVALS.txt to set different
    parameters, \n");
    printf("including COMM PORT NUMBER TO BE USED->
    1st entry in IVALS.txt \n \n");
    exit(EXIT_FAILURE);
  }
  else
  {
    printf("Execution will continue, using COMM PORT
    %c \n", commNum);
  }

  commStr[7] = commNum;
```

// OPEN FILE "RTATT2IMUOUT.txt"–
// RESULTS WILL BE WRITTEN TO THIS FILE AS C S V

```
  fp = fopen("RTATT2IMUOUT.txt", "w"); // "w"=re-write,
  "a"=addline

  // ++++++++++++++++++++++++++++++++++++
```

```
// INITIALIZATION OF KF MATRICES
MAKESCALEDI4(PAfact, PA); // Scaling PA
MAKESCALEDI4(Qfact, Q); // Scaling Q
MAKESCALEDI4(Hfact, H); // Scaling H
MAKESCALEDI4(Rfact, R); // Scaling R

xA.x = 1; // Initialize xA to [1, 0, 0, 0]T
xA.y = 0;
xA.z = 0;
xA.w = 0;

 com_handle = openAndSetupComPort(commStr);
 if (com_handle == INVALID_HANDLE_VALUE) {
 printf("comm port open failed\n");
 }
 if (setupStreaming(com_handle)) {

 printf("Streaming set up failed\n");
 }
 printf("\n \n Press <Enter> to tare the sensor and
 start streaming . . . \n");
 getchar(); getchar();

 write_tare_bytes[0] = TSS_START_BYTE;
 write_tare_bytes[1] = TSS_TARE_CURRENT_ORIENTATION;
 write_tare_bytes[2] = createChecksum(&write_tare_
 bytes[1], 1);
 if (!WriteFile(com_handle, write_tare_bytes,
 sizeof(write_tare_bytes), &bytes_written, 0)) {
 printf("Error to tare the sensor\n");
 }

 printf("\n STREAMING STARTED—IT WILL CONTINUE FOR
 %d SECONDS\n", TIMERUN);
 // With parameterless wired commands the command byte
 // will be the same as the checksum
 write_stream_bytes[0] = TSS_START_BYTE;
 write_stream_bytes[1] = TSS_START_STREAMING;
 write_stream_bytes[2] = TSS_START_STREAMING;
```

```
// Write the bytes to the serial
if (!WriteFile(com_handle, write_stream_bytes,
sizeof(write_stream_bytes), &bytes_written, 0)) {
printf("Error writing to port\n");
//return 3;
}
QueryPerformanceCounter(&start_time); // Retrieves
the current value of the high-resolution performance
counter.
QueryPerformanceCounter(&current_time); // Retrieves
the current value of the high-resolution performance
counter.
  // while loop runs for as many seconds as assigned in
  // STREAM_DURATION = TIMERUN
  while ((float)STREAM_DURATION > ((current_time.
  QuadPart−start_time.QuadPart) * 1.0f/frequency.
  QuadPart))
{
QueryPerformanceCounter(&current_time); // Retrieves
the current value of the high-resolution performance
counter
   // Read the bytes returned from the serial
   if (!ReadFile(com_handle, &batch_data,
   sizeof(batch_data), &bytes_read, 0)) {
   printf("Error reading from port\n");
   }
   if (bytes_read != sizeof(batch_data)) {
   continue;
   }
   // The data must be flipped to little endian to be read correctly
   for (i = 0; i < sizeof(batch_data.Data)/
   sizeof(float); i++) {
   endian_swap_32((unsigned int*)&batch_data.Data[i]);
   }
   // Calculate current time
   CurrentTime = (current_time.QuadPart−start_time.
   QuadPart) * 1.0f/frequency.QuadPart;
   // Parsing Sensor Data
   Stillness = batch_data.Data[0];
   Gy.x = batch_data.Data[1];
   Gy.y = batch_data.Data[2];
   Gy.z = batch_data.Data[3];
```

```
    Ac.x = batch_data.Data[4];
    Ac.y = batch_data.Data[5];
    Ac.z = batch_data.Data[6];
    IMUquat.x = batch_data.Data[7];
    IMUquat.y = batch_data.Data[8];
    IMUquat.z = batch_data.Data[9];
    IMUquat.w = batch_data.Data[10];
    Compass.x = batch_data.Data[11];
    Compass.y = batch_data.Data[12];
    Compass.z = batch_data.Data[13];
    //Confidence = batch_data.Data[14];
// *** ********* START -Read Gyro, accel to KF variables
    a = (-1) * Gy.z;
    b = (-1) * Gy.x;
    c = Gy.y;
    dt2 = dt / 2;
    // Forming Matrix F, as F = I4 + (dt/2) * OMEGA_MATRIX
    F[0][0] = 1;
    F[0][1] = (-1) * dt2 * a;
    F[0][2] = (-1) * dt2 * b;
    F[0][3] = (-1) * dt2 * c;
    F[1][0] = (dt2)*a;
    F[1][1] = 1;
    F[1][2] = (dt2)*c;
    F[1][3] = (-1) * dt2 * b;
    F[2][0] = (dt2)*b;
    F[2][1] = (-1) * dt2 * c;
    F[2][2] = 1;
    F[2][3] = (dt2)*a;
    F[3][0] = (dt2)*c;
    F[3][1] = (dt2)*b;
    F[3][2] = (-1) * dt2 * a;
    F[3][3] = 1;
    Acmag = sqrt((Ac.x * Ac.x) + (Ac.y * Ac.y) +
    (Ac.z * Ac.z));
    AcxN = Ac.z/Acmag;
    AcyN = Ac.x/Acmag;
    //      AczN = Ac.y/Acmag;
    tht = asin(AcxN); // Calculate theta
    phi = asin(((-1) * AcyN)/(cos(tht)));
    psi = 0;
    th2 = tht / 2;
```

```
ph2 = phi / 2;
ps2 = psi / 2;
// LODING OF ACCELERATIONS TO Z VECTOR
z.x = cos(ph2) * cos(th2) * cos(ps2) + sin(ph2) *
sin(th2) * sin(ps2);
z.y = sin(ph2) * cos(th2) * cos(ps2)—cos(ph2) *
sin(th2) * sin(ps2);
z.z = cos(ph2) * sin(th2) * cos(ps2) + sin(ph2) *
cos(th2) * sin(ps2);
z.w = cos(ph2) * cos(th2) * sin(ps2)—sin(ph2) *
sin(th2) * cos(ps2);
// *** ********* END -Read Gyro, accel to KF vars
// KF PROCESSING ########
// (Uses generic 4x4 matrices ML, MO, MP, MT for TEMP. storage)
// ----- PREDICTION EQUATIONS
xB = M44V41(F, xA); // EQ 12.1: xB = F * XA
(There is no u vector in this application)
FPFT(F, PA, ML); // Term F * P * FT of EQ. 12.2
is calculated and stored in ML
M44PLUSM44(ML, Q, PB); // Completing EQ. 12.2.
Result stored in PB
// ----- CORRECTION EQUATIONS
SI4SI4SI4(H, PB, H, MN); // PROD HPBHT in MN—Can
use H instead of HT b/c H is symmetric
M44PLUSM44(MN, R, MO); // (HPBHT + R) computed
and held in MO
SI4INV(MO, MS); // INVERSE OF(HPBHT + R) computed
and held in MS
SI4SI4SI4(PB, H, MS, KG); // Completing EQ. 12.3,
use H instead of HT because its symm.
xA = V41PLUSV41(xB, (M44V41(KG, (V41MINUSV41(z,
(M44V41(H, xB))))))); // EQ. 12.4
SI4SI4SI4(KG, H, PB, MT); // The term (Kg * H *
PB) is computed. Result in MT
M44MINUSM44(PB, MT, PA); // COMPLETING EQ. 12.5.
RESULT IN PA
// END OF KF PROCESSING ########
// CONVERSION TO THETA, PSI, PHI OF RESULTING
QUATERNION
fltphi = atan2((2 * (xA.z * xA.w + xA.x * xA.y)),
(1-2 * (pow(xA.y, 2) + pow(xA.z, 2)))) * r2d;
```

```
    fltthtt = (-1) * (asin(2 * (xA.y * xA.w—xA.x *
    xA.z))) * r2d;
    fltpsi = atan2((2 * (xA.y * xA.z + xA.x * xA.w)),
    (1-2 * (pow(xA.z, 2) + pow(xA.w, 2)))) * r2d;
   fprintf(fp, "%f,%f,%f,%f,%f,%f,%f \n", CurrentTime,
   (phi * r2d), (tht * r2d), (psi * r2d), fltphi,
   fltthtt, fltpsi);
    sample_count++;
  } // End of while loop—Streaming went on for STREAM_
  DURATION = TIMERUN seconds
   CloseHandle(com_handle);
  fclose(fp);
  printf("\nSample Count = %u samples in %.4f
  seconds\n", sample_count, CurrentTime);
  printf("Effective Sampling Interval, dt = %f seconds
  \n", ( CurrentTime/(float)sample_count));
  printf("\nPort is closed");
  // END of PROGRAM
  printf("\nFinished! press <Enter> to terminate");
  getchar();
  return 0;
}
```

//##
// ADDITIONAL FUNCTIONS FOR MATRIX, VECTOR OPERATIONS
// IN THE KALMAN FILTER IMPLEMENTATION
//##
//FUNCTION M44V41

```
//MULTPLIES a GRAL 4x4 MTX times a 4X1 VECTOR
vec4f M44V41(float M[4][4], vec4f V)
{
  vec4f R;
  R.x = (M[0][0] * V.x) + (M[0][1] * V.y) + (M[0][2] *
  V.z) + (M[0][3] * V.w);
  R.y = (M[1][0] * V.x) + (M[1][1] * V.y) + (M[1][2] *
  V.z) + (M[1][3] * V.w);
  R.z = (M[2][0] * V.x) + (M[2][1] * V.y) + (M[2][2] *
  V.z) + (M[2][3] * V.w);
  R.w = (M[3][0] * V.x) + (M[3][1] * V.y) + (M[3][2] *
  V.z) + (M[3][3] * V.w);
```

```
  return R;
  }
```

//FUNCTION M44PLUSM44

```
//ADDS TWO 4x4 General matrices
void M44PLUSM44(float M1[4][4], float M2[4][4], float
MR[4][4])
{
  int i, j;
  for (i = 0; i <= 3; i++)
  {
    for (j = 0; j <= 3; j++)
    {
      MR[i][j] = M1[i][j] + M2[i][j];
    }
}
}
```

//FUNCTION M44MINUSM44

```
//SUBSTRACTS M1 – M2, both 4 x 4
void M44MINUSM44(float M1[4][4], float M2[4][4], float
MR[4][4])
{
  int i, j;
  for (i = 0; i <= 3; i++)
  {
    for (j = 0; j <= 3; j++)
    {
      MR[i][j] = M1[i][j]–M2[i][j];
    }
}
}
```

//FUNCTION V41PLUSV41

```
// ADDS V1 + V2 , BOTH 4x1 VECTORS
vec4f V41PLUSV41(vec4f V1, vec4f V2)
{
  vec4f R;
  R.x = V1.x + V2.x;
  R.y = V1.y + V2.y;
  R.z = V1.z + V2.z;
  R.w = V1.w + V2.w;
  return R;
}
```

//FUNCTION V41MINUSV41

```
// SUBTRACTS V1 - V2 , BOTH 4x1 VECTORS
vec4f V41MINUSV41(vec4f V1, vec4f V2)
{
  vec4f R;
  R.x = V1.x-V2.x;
  R.y = V1.y-V2.y;
  R.z = V1.z-V2.z;
  R.w = V1.w-V2.w;
  return R;
}
```

//FUNCTION MAKESCALEDI4

```
// CREATES SCALED I4 IN M1, APPLYING ScaleFactor
void MAKESCALEDI4(float ScaleFactor, float M1[4][4])
{
  int i, j;
  for (i = 0; i <= 3; i++) // writing zeros on all
  entries
  {
    for (j = 0; j <= 3; j++)
    {
    M1[i][j] = 0;
    }
  }
  M1[0][0] = ScaleFactor; // WRITING ScaleFactor ON
  THE 4 DIAGONAL ELEMENTS
  M1[1][1] = ScaleFactor;
  M1[2][2] = ScaleFactor;
  M1[3][3] = ScaleFactor;
}
```

//FUNCTION SI4SI4

```
//MULTPLIES TWO 4x4 Scaled Id. matrices
void SI4SI4(float M1[4][4], float M2[4][4], float
MR[4][4])
{
  float ScalarRes = 0;
  int i, j;
  for (i = 0; i <= 3; i++) // writing zeros on all
  entries
  {
    for (j = 0; j <= 3; j++)
```

```
    {
    MR[i][j] = 0;
    }
   }
   ScalarRes = M1[0][0] * M2[0][0]; // calculate scalar
   for diag. of result
   MR[0][0] = ScalarRes; //write the scalar for result
   in the 4 diag entries of result
   MR[1][1] = ScalarRes;
   MR[2][2] = ScalarRes;
   MR[3][3] = ScalarRes;
}
```

//FUNCTION SI4SI4SI4

```
//MULTPLIES THREE 4x4 Scaled Id. matrices
void SI4SI4SI4(float M1[4][4], float M2[4][4], float
M3[4][4], float MR[4][4])
{
  float ScalarRes = 0;
  int i, j;
  for (i = 0; i <= 3; i++) // writing zeros on all
  entries
  {
    for (j = 0; j <= 3; j++)
    {
    MR[i][j] = 0;
    }
  }
  ScalarRes = M1[0][0] * M2[0][0] * M3[0][0]; //
  calculate scalar for diag. of result
  MR[0][0] = ScalarRes; //write the scalar for result
  in the 4 diag entries of result
  MR[1][1] = ScalarRes;
  MR[2][2] = ScalarRes;
  MR[3][3] = ScalarRes;
}
```

//FUNCTION SI4V41

```
//MULTPLIES a scaled I4 times a 4X1 VECTOR
vec4f SI4V41(float SI[4][4], vec4f V)
{
  vec4f R;
  float ScaleF;
```

```
  ScaleF = SI[0][0];
  R.x = ScaleF * V.x;
  R.y = ScaleF * V.y;
  R.z = ScaleF * V.z;
  R.w = ScaleF * V.w;
  return R;
}
```

//FUNCTION SI4INV

```
//GETS INVERSE OF 4x4 Scaled Id. matrix
void SI4INV(float M1[4][4], float MR[4][4])
{
  int i, j;
  for (i = 0; i <= 3; i++) // writing zeros on all
  entries
  {
    for (j = 0; j <= 3; j++)
    {
      MR[i][j] = 0;
    }
  }
  MR[0][0] = 1 / (M1[0][0]); //write the inv for each
  of the 4 diag entries of result
  MR[1][1] = 1 / (M1[1][1]);
  MR[2][2] = 1 / (M1[2][2]);
  MR[3][3] = 1 / (M1[3][3]);
}
```

//FUNCTION FPFT

```
//IMPLEMENTS F*P*FT ASSUMING P is Scaled ID
void FPFT(float F[4][4], float P[4][4], float MR[4]
[4])
{
  int i, j;
  float ScalarRes, a2, b2, c2;
  a2 = (F[0][1] * F[0][1]);
  b2 = (F[0][2] * F[0][2]);
  c2 = (F[0][2] * F[0][2]);
  ScalarRes = P[0][0] * (a2 + b2 + c2 + 1);
  // initialize the result matrix with all 16 zeros
  for (i = 0; i <= 3; i++) // writing zeros on all
  entries
  {
```

```
    for (j = 0; j <= 3; j++)
    {
      MR[i][j] = 0;
    }
}
  MR[0][0] = ScalarRes; //write the scalar for result
  in the 4 diag entries of result
  MR[1][1] = ScalarRes;
  MR[2][2] = ScalarRes;
  MR[3][3] = ScalarRes;
}
```

//FUNCTION PRNM44 — ONLY USEFUL FOR DEBUGGING

```
//PRINTS a GENERAL 4 x 4 FLOAT MATRIX TO CONSOLE
void PRNM44(float M[4][4])
{
  printf(" \n");
  printf("%f, %f, %f, %f\n", M[0][0], M[0][1], M[0]
  [2], M[0][3]);
  printf("%f, %f, %f, %f\n", M[1][0], M[1][1], M[1]
  [2], M[1][3]);
  printf("%f, %f, %f, %f\n", M[2][0], M[2][1], M[2]
  [2], M[2][3]);
  printf("%f, %f, %f, %f\n", M[3][0], M[3][1], M[3]
  [2], M[3][3]);
}
```

//FUNCTION PRNV41 — ONLY USEFUL FOR DEBUGGING

```
//PRINTS (TO CONSOLE) a GENERAL 4 x 1 FLOAT VECTOR TO
CONSOLE
void PRNV41(vec4f V)
{
  printf(" \n");
  printf("%f \n", V.x);
  printf("%f \n", V.y);
  printf("%f \n", V.z);
  printf("%f \n", V.w);
}
```

```
// + + + + + + FILE RTATT2IMU.h + + + + + + + + + + + +
#include <stdio.h>
#include <Windows.h>
#include "yei_threespace_basic_utils.h"
// For a full list of streamable commands refer to "Wired Streaming
Mode" section
// in the 3-Space Manual of the sensor
#define TSS_TARE_CURRENT_ORIENTATION 0x60
#define TSS_SET_AXIS_DIRECTION 0x74
#define TSS_GET_TARED_ORIENTATION_AS_QUAT 0x00
#define TSS_GET_TARED_ORIENTATION_AS_PYR 0x01
#define TSS_GET_NORTH_AND_GRAVITY 0x0A
#define TSS_GET_FORWARD_AND_DOWN 0x0B
#define TSS_GET_UNTARED_2_VECTOR_IN_SENSOR_FRAME 0x0C
#define TSS_UPDATE_CURRENT_TIMESTAMP 0x5F
#define TSS_GET_TARED_ORIENTATION_AS_AXI 0x03
#define TSS_READ_TEMP_FAH 0x2C
#define TSS_SET_PEDESTRIAN_TRACKING 0x34
#define TSS_SET_PEDESTRIAN_TRACKING_ON_OFF 0x00
#define TSS_GET_PEDESTRIAN_TRACKING 0x35
#define TSS_GET_CONFIDENCE_VALUE 0x15
#define TSS_GET_NORMALIZED_UNIT_VEC_ACCELEROMETER 0x22
#define TSS_GET_CORRECTED_LINEAR_ACC_AND_GRAVITY 0x27
#define TSS_GET_CORRECTED_LINEAR_ACC_ONLY 0x29
#define TSS_GET_RAD_PER_SEC_GYROSCOPE 0x26
#define TSS_GET_NORMALIZED_COMPASS 0x23
#define TSS_GET_CORRECTED_COMPASS 0x28
#define TSS_GET_RAW_COMPASS_DATA 0x43
#define TSS_GET_GYRO_CALIBRATE_COEFFS 0xA4
#define TSS_NULL 0xff // No command use to fill the
empty slots in "set stream slots"
// For a full list of commands refer to the 3-Space Manual of the sensor
#define TSS_SET_STREAMING_SLOTS 0x50
#define TSS_SET_STREAMING_TIMING 0x52
#define TSS_START_STREAMING 0x55
#define TSS_STOP_STREAMING 0x56
#define THRSHLD 0.75 //Stillness Threshold
#define N 50 // Number of samples used to determine
Gyro Bias
// Stream data structures must be packed else they
will not properly work
```

```
#pragma pack(push,1)
typedef struct Batch_Data{
 float Data[14];
} Batch_Data;
#pragma pack(pop)
typedef struct vec3f
{
 float x;
 float y;
 float z;
}
vec3f;
typedef struct vec4f
{
 float x;
 float y;
 float z;
   float w;
}
vec4f;
// Streaming mode require the streaming slots and streaming timing
// being setup prior to start streaming
int setupStreaming(HANDLE com_handle)
{
 DWORD bytes_written;
 unsigned char write_slot_bytes[11]; // start byte,
 command, data(8), checksum
 unsigned char write_timing_bytes[15]; // start byte,
 command, data(12), checksum
   unsigned char write_pedestrian_bytes[5];
 unsigned int interval;
 unsigned int duration;
 unsigned int delay;
 printf(">> SETTING STREAMING SLOTS VIA >> \n");
   // SET STREAMING SLOTS //
 // There are 8 streaming slots available for use, and each one can hold one
 // of the streamable commands.
 // Unused slots should be filled with 0xff so that they will output nothing.
 write_slot_bytes[0]= TSS_START_BYTE;
 write_slot_bytes[1]= TSS_SET_STREAMING_SLOTS;
 write_slot_bytes[2]= TSS_READ_TEMP_FAH; // stream slot0
```

```
write_slot_bytes[3]= TSS_GET_RAD_PER_SEC_GYROSCOPE;
// stream slot1
write_slot_bytes[4]= TSS_GET_CORRECTED_LINEAR_ACC_
AND_GRAVITY; // stream slot2
write_slot_bytes[5]= TSS_GET_TARED_ORIENTATION_AS_
QUAT; // stream slot3
write_slot_bytes[6]= TSS_GET_NORMALIZED_COMPASS; //
stream slot4
write_slot_bytes[7]= TSS_NULL; // stream slot5
write_slot_bytes[8]= TSS_NULL; // stream slot6
write_slot_bytes[9]= TSS_NULL; // stream slot7
write_slot_bytes[10]= createChecksum(&write_slot_
bytes[1], 8+1);
// Write the bytes to the serial
if(!WriteFile(com_handle, write_slot_bytes,
sizeof(write_slot_bytes), &bytes_written, 0)){
printf("Error writing to port\n");
return 1;
}
// SET STREAMING TIMING //
// Interval determines how often the streaming session will output
data
// from the requested commands
// An interval of 0 will output data at the max filter rate
interval= 100000; // microseconds
// Duration determines how long the streaming session will run for
// A duration of 0xffffffff will have the streaming session run till
// the stop stream command is called
//duration= STREAM_DURATION*1000000; // microseconds
  duration= 0xffffffff; // microseconds
// Delay determines how long the sensor should wait after a start
command
// is issued to actually begin streaming
delay= 100000; //microseconds
//The data must be flipped to big endian before sending to sensor
endian_swap_32((unsigned int *)&interval);
endian_swap_32((unsigned int *)&duration);
endian_swap_32((unsigned int *)&delay);
write_timing_bytes[0]= TSS_START_BYTE;
write_timing_bytes[1]= TSS_SET_STREAMING_TIMING;
```

```
    memcpy(&write_timing_bytes[2], &interval,
    sizeof(interval));
    memcpy(&write_timing_bytes[6], &duration,
    sizeof(duration));
    memcpy(&write_timing_bytes[10], &delay,
    sizeof(delay));
    write_timing_bytes[sizeof(write_timing_bytes)-1]=
    createChecksum(&write_timing_bytes[1], 12+1);
    // Write the bytes to the serial
    if(!WriteFile(com_handle, write_timing_bytes,
    sizeof(write_timing_bytes), &bytes_written, 0)){
      printf("Error writing to port\n");
      return 2;
    }
    return 0;
}
```

References

Aggarwal, P., Z. Syed, and A. Noureldin. 2010. *MEMS-Based Integrated Navigation*: Artech House, Boston, MA.

Ahrendt, Peter. 2005. "The Multivariate Gaussian Probability Distribution." Technical University of Denmark, Technical Report.

Anderson, B.D.O., and J.B. Moore. 2012. *Optimal Filtering*: Dover Publications, Mineola, NY.

Ayman, S. 2016. *Sensor Fusion Using Kalman Filter for a Quadrotor-Attitude Estimation: Basics, Concepts, Modelling, MATLAB Code and Experimental Validation*: LAP Lambert Academic Publishing, Saarbrucken, Germany.

Bishop, Gary, and Greg Welch. 2001. "An Introduction to the Kalman Filter." *Proceedings of SIGGRAPH, Course* 8 (27599–23175): 41.

Bromiley, P. A. 2003. "Products and Convolutions of Gaussian Distributions." Medical School, University of Manchester, Manchester, Technical Report 3.

Brown, R.G., and P.Y.C. Hwang. 2012. *Introduction to Random Signals and Applied Kalman Filtering with MATLAB Exercises, 4th Edition*: John Wiley & Sons, Incorporated, Hoboken, NJ.

Candy, J.V. 2009. *Bayesian Signal Processing: Classical, Modern and Particle Filtering Methods*: John Wiley & Sons, Incorporated, Hoboken, NJ.

Catlin, D.E. 1988. *Estimation, Control, and the Discrete Kalman Filter*: Springer, New York, NY.

Childers, D.G. 1997. *Probability and Random Processes: Using MATLAB with Applications to Continuous and Discrete Time Systems*: Irwin (McGraw-Hill), New York, NY.

Cox, B., and A. Cohen. 2014. *Human Universe*: William Collins, (Harper Collins), New York, NY.

Drake, S. 1978. *Galileo at Work: His Scientific Biography*: University of Chicago Press, Chicago, IL.

Farrell, J.A. 2008. *Aided Navigation: GPS with High Rate Sensors*: McGraw-Hill Higher Education, New York, NY.

Field, A. 2013. *Discovering Statistics Using IBM SPSS Statistics*: Sage Publications, London, England.

Gales, M.J.F., and S.S. Airey. 2006. "Product of Gaussians for Speech Recognition." *Computer Speech & Language* 20 (1): 22–40. doi:10.1016/j.csl.2004.12.002.

Gelb, A., J.F. Kasper, Analytic Sciences Corporation, Technical Staff, and Analytic Sciences Corporation. 1974. *Applied Optimal Estimation*: M.I.T. Press, Cambridge, MA.

Grewal, M.S., and A.P. Andrews. 2008. *Kalman Filtering: Theory and Practice Using MATLAB*: John Wiley & Sons, Incorporated, Hoboken, NJ.

Haug, A.J. 2012. *Bayesian Estimation and Tracking: A Practical Guide*: John Wiley & Sons, Incorporated, Hoboken, NJ.

Hayes, M.H. 1996. *Statistical Digital Signal Processing and Modeling*: John Wiley & Sons, Incorporated, Hoboken, NJ.

Holman, J.P. 2001. *Experimental Methods for Engineers*: McGraw-Hill Higher Education, New York, NY.

Johnson, R.A., and D.W. Wichern. 2002. *Applied Multivariate Statistical Analysis*: Prentice Hall, Upper Saddle River, NJ.

Kalman, R.E. 1960. "A New Approach to Linear Filtering and Prediction Problems." *Journal of Basic Engineering* 82 (1): 35–45. doi:10.1115/1.3662552.

Kovvali, N., M. Banavar, and A. Spanias. 2013. *An Introduction to Kalman Filtering with MATLAB Examples*: Morgan & Claypool Publishers, San Rafael, CA.

Kuipers, J.B. 2002. *Quaternions and Rotation Sequences: A Primer with Applications to Orbits, Aerospace, and Virtual Reality*: Princeton University Press, Princeton, NJ.

Lathi, B.P. 1998. *Signal Processing and Linear Systems*: Berkeley-Cambridge Press, Carmichael, CA.

Maybeck, P.S. 1979. *Stochastic Models, Estimation and Control*: Academic Press, New York, NY.

McClellan, J.H., R.W. Schafer, and M.A. Yoder. 2003. *Signal Processing First*: Pearson Education International, Upper Saddle River, NJ.

McClellan, J.H., R.W. Schafer, and M.A. Yoder. 2016. *DSP First*: Pearson, Upper Saddle River, NJ.

McCool, J.I. 2012. *Using the Weibull Distribution: Reliability, Modeling, and Inference*: John Wiley & Sons, Incorporated, Hoboken, NJ.

Mix, D.F. 1995. *Random Signal Processing*: Prentice Hall, Englewood Cliffs, NJ.

O'Connell, R., J.B. Orris, and B. Bowerman. 2011. *Essentials of Business Statistics*: McGraw-Hill Higher Education, New York, NY.

Papoulis, A., and S.U. Pillai. 2002. *Probability, Random Variables, and Stochastic Processes*: McGraw-Hill Higher Education, New York, NY.

Peebles, P.Z. 2001. *Probability, Random Variables, and Random Signal Principles*: McGraw-Hill Higher Education, New York, NY.

Rizzoni, G. 2004. *Principles and Applications of Electrical Engineering*: McGraw-Hill Higher Education, New York, NY.

Serra, G.L. 2018. *Kalman Filters: Theory for Advanced Applications*: IntechOpen, Rijeka, Croatia.

Shoemake, Ken. 1985. "Animating Rotation with Quaternion Curves." *ACM SIGGRAPH Computer Graphics* 19 (3): 245–254

Simon, Dan. 2001. "Kalman Filtering." *Embedded Systems Programming* 14 (6): 72–79.

Snider, A.D. 2017. *Random Processes for Engineers: A Primer*: CRC Press, Taylor & Francis Group, Boca Raton, FL.

Spagnolini, U. 2018. *Statistical Signal Processing in Engineering*: John Wiley & Sons, Incorporated, Hoboken, NJ.

Stark, H., and J.W. Woods. 2002. *Probability and Random Processes with Applications to Signal Processing*: Prentice Hall, Englewood Cliffs, NJ.

Stengel, R.F. 1994. *Optimal Control and Estimation*: Dover Publications, New York, NY.

Stone, J.V. 2013. *Bayes' Rule: A Tutorial Introduction to Bayesian Analysis*: Sebtel Press, Sheffield.

Stone, J.V. 2015. *Bayes' Rule with MATLAB: A Tutorial Introduction to Bayesian Analysis*: Sebtel Press, Sheffield.

Stone, J.V. 2016a. *Bayes' Rule with Python: A Tutorial Introduction to Bayesian Analysis*: Sebtel Press, Sheffield.

Stone, J.V. 2016b. *Bayes' Rule with R: A Tutorial Introduction to Bayesian Analysis*: Sebtel Press, Sheffield.

Stone, J. V. 2019. "An Introduction to Bayes Rule (Chapter 1, Online)." [Web Page], accessed 11/28/2019. http://jim-stone.staff.shef.ac.uk/BookBayes2012/HTML_BayesRulev5E-bookHTMLFiles/ops/xhtml/ch01BayesJVSone.html.

Titterton, D.H., and J.L. Weston. 2004. *Strapdown Inertial Navigation Technology*: American Institute of Aeronautics and Astronautics, Reston, VA.

Wertz, J.R. 1980. *Spacecraft Attitude Determination and Control*: Reidel, Dordrecht, Holland.

Yates, R.D., and D.J. Goodman. 2014. *Probability and Stochastic Processes: A Friendly Introduction for Electrical and Computer Engineers*: John Wiley & Sons, Incorporated, Hoboken, NJ.

Yun, X., E. R. Bachmann, and R. B. McGhee. 2008. "A Simplified Quaternion-Based Algorithm for Orientation Estimation from Earth Gravity and Magnetic Field Measurements." *IEEE Transactions on Instrumentation and Measurement* 57 (3): 638–650. doi:10.1109/tim.2007.911646.

Zarchan, P., H. Musoff, F.K. Lu, American Institute of Aeronautics, and Astronautics. 2009. *Fundamentals of Kalman Filtering: A Practical Approach*: American Institute of Aeronautics and Astronautics, Reston, VA.

Index

Note: Page numbers in *italic* indicate a figure and page numbers in **bold** indicate a table on the corresponding page.

S

V

W

Y

Z

Printed in the United States
by Baker & Taylor Publisher Services

Printed in the United States
by Baker & Taylor Publisher Services